Nanostructured
Titanium Dioxide
Materials

Properties, Preparation
and Applications

Nanostructured Titanium Dioxide Materials

Properties, Preparation and Applications

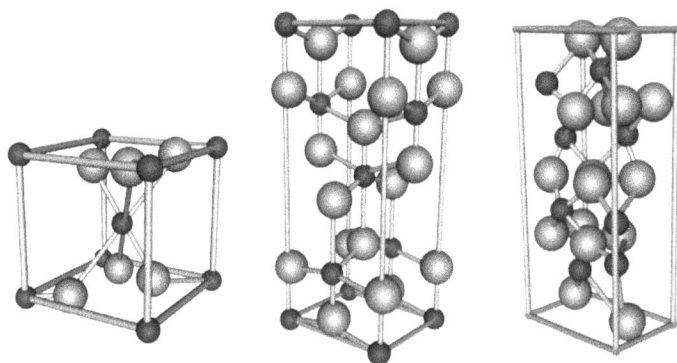

Alireza Khataee

University of Tabriz, Iran

G Ali Mansoori

University of Illinois at Chicago, USA

World Scientific

NEW JERSEY · LONDON · SINGAPORE · BEIJING · SHANGHAI · HONG KONG · TAIPEI · CHENNAI

Published by

World Scientific Publishing Co. Pte. Ltd.

5 Toh Tuck Link, Singapore 596224

USA office: 27 Warren Street, Suite 401-402, Hackensack, NJ 07601

UK office: 57 Shelton Street, Covent Garden, London WC2H 9HE

British Library Cataloguing-in-Publication Data
A catalogue record for this book is available from the British Library.

ISBN-13 978-981-4374-72-9
ISBN-10 981-4374-72-5

Printed by FuIsland Offset Printing (S) Pte Ltd Singapore

Brief Summary

In the past decade, research and development in the area of synthesis and application of different nanostructured titanium dioxide (nanowires, nanotubes, nanfibers and nanoparticles) have become tremendous. This book briefly describes the properties, production, modification and applications of nanostructured titanium dioxide. Special emphasis is placed on photocatalytic activity as well as on some requirements for efficient photocatalysts. The physicochemical properties of nanostructured titanium dioxide are highlighted and the links between properties and applications are described. The preparation of TiO_2 nanomaterials, including nanoparticles, nanorods, nanowires, nanosheets, nanofibers and nanotubes are primarily categorized with the relevant preparation method (e. g. sol–gel and hydrothermal processes). Examples of early applications of nanostructured titanium dioxide in dye–sensitized solar cells, hydrogen production and storage, sensors, rechargeable batteries, self–cleaning and antibacterial surfaces electrocatalysis and photocatalytic cancer treatment are then reviewed. Since many applications of TiO_2 nanomaterials are closely related to their optical properties, this book presents a section on the research related to the modifications of the optical properties of TiO_2 nanomaterials. TiO_2 nanomaterials normally are transparent in the visible light region. By doping, it is possible to improve the optical sensitivity and activity of TiO_2 nanomaterials in the visible light region. Photocatalytic removal of various pollutants using pure TiO_2 nanomaterials, TiO_2–based nanoclays and non–metal doped nanostructured TiO_2 are also discussed. Finally, we describe immobilization methods of TiO_2 nanomaterials on different substrates (e.g. glass, ceramic, stone, cement, zeolites, metallic and metal oxide materials and polymer substrates).

Keywords: Titanium dioxide, Titanate nanotubes, Nanoparticles, Nanosheets, Nanofibers, NS–TiO_2, Sol–gel process, Nanoclays, Doped–TiO_2, Hydrothermal process, Photocatalysis, Electrocatalysis, Solar cell, Lithium batteries, Antibacterial surfaces, Self–cleaning surfaces, Photocatalytic cancer treatment, H_2 production, Environmental remediation, Immobilized TiO_2.

Contents

Chapter 1

Introduction

Titanium Dioxide (TiO_2) has a wide range of applications. Since its commercial production in the early twentieth century, it is used as a pigment in paints, coatings, sunscreens, ointments and toothpaste. TiO_2 is considered a "quality–of–life" product with demand affected by gross domestic product in various regions of the world. Titanium dioxide pigments are inorganic chemical products used for imparting whiteness, brightness and opacity to a diverse range of applications and end–use markets. TiO_2 as a pigment derives value from its whitening properties and opacifying ability (commonly referred to as *hiding power*). As a result of TiO_2's high refractive index rating, it can provide more hiding power than any other commercially available white pigment. Titanium dioxide is obtained from a variety of ores that contain ilmenite, rutile, anatase and leucoxene, which are mined from deposits located throughout the world. The commercial production of this pigment started in the early twentieth century during the investigation of ways to convert ilmenite to iron or titanium–iron alloys. The first industrial production of TiO_2 started in 1918 in Norway, the United State and Germany. Crystals of titanium dioxide exist in three crystalline forms: Rutile, Anatase and Brookite (see Figures 1 and 2). Only anatase and rutile forms have good pigmentary properties. However, rutile is more thermally stable than anatase. Most titanium dioxide pigments, either as the rutile or the anatase form, are produced from titanium mineral concentrates through a chloride or sulfate process [1–3].

The purpose of this report is to present and discuss properties, production, modification and applications of nanostructured titanium dioxide ($NS–TiO_2$). With the advent of nanotechnology, $NS–TiO_2$ has found a great deal of applications. Nanotechnology is a growing and cutting edge technology that has influenced many fields of research and development areas such as biology, chemistry, material science, medicine and physics. With the inception of nanoscience and nanotechnology, nanoscale materials like *NS–TiO₂* have received significant attention. The typical dimension of *NS–TiO₂* is less than 100 nm, which makes it attractive for numerous applications in different fields. *NS–TiO₂* is known to have such *golden* properties as abundance and potentially low cost compared to other nanomaterials. *NS–TiO₂* materials include spheroidal nanocrystallite and nanoparticles along with elongated nanotubes, nanosheets and nanofibers [4–7].

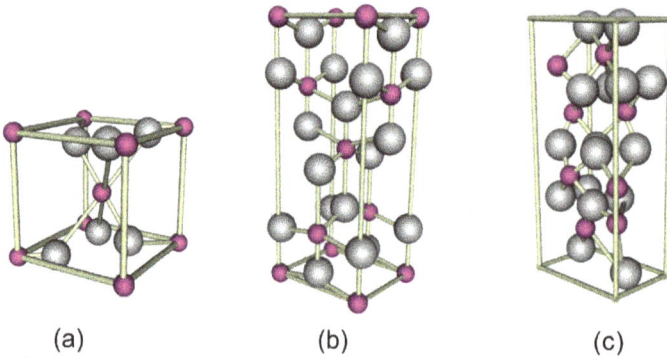

Figure 1. Unit cells of (A) rutile, (B) anatase and (C) brookite. Grey and red spheres represent oxygen and titanium, respectively.

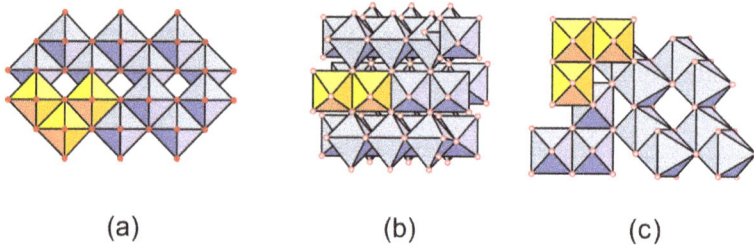

Figure 2. Crystalline structure of (A) anatase, (B) rutile and (C) brookite. (Adapted from Khataee and Kasiri [1] with permission from publisher, Elsevier. License Number: 2627060102098).

The field of nanotechnology has generated a great deal of interest primarily because in nano size–scaled, materials have numerous new and innate properties. These size–dependent properties include new phase transition behavior, peculiar thermal and mechanical properties, interesting surface activity and reactivity (catalysis) and unusual optical, electrical and magnetic characteristics [8–13].

Among the unique properties of nanomaterials, the movement of electrons and holes in semiconductor nanomaterials is primarily governed by quantum confinement. Nanomaterials transport properties related to phonons and photons are largely affected by their nano size and geometry. The specific surface area and surface–to–volume ratio increase dramatically as the size of a material decreases. The high surface area brought about by small particle size is beneficial to many TiO_2–based devices. It facilitates the reaction/interaction between the device and the interacting media, which mainly occurs on the surface or at the interface. Thus, the performance of TiO_2–based devices is largely influenced by the size of TiO_2 building units [14, 15].

A number of reviews and reports on different aspects of titanium dioxide, including its properties, preparation, modification and application, have been published [16–22]. Fox and Dulay [23] briefly discussed the cogent features of the irradiated TiO_2 surface and provided an overview of typical photocatalytic reactions observed on heterogeneous dispersions of semiconductors. They also described experiments that help to define the mechanism of such photocatalysis. Yates *et al.* [24] analysed some of the operating principles of heterogeneous TiO_2 photocatalysis. They examined the electronic excitation processes in the TiO_2 molecule. The electronic interactions between the adsorbate molecule and the catalyst substrate were discussed in terms of the catalyzed or sensitized photoreactions. The research group also summarized thermal and photocatalytic studies of TiO_2 with emphasis on the common characteristics and fundamental principles of TiO_2–based photocatalytic processes. In a recent review article, Yates and Thompson discussed the surface science of the photoactivation of TiO_2 as a new photochemical process [25]. Hoffmann *et al.* [26] formulated a comprehensive review of the environmental applications of semiconductor photocatalysis. Hoffmann and co–workers provided an overview of some of the underlying principles governing semiconductor photocatalysis and reviewed literature in terms of TiO_2's potential applications in the environmental control technology. Fujishima *et al.* developed two review articles regarding photocatalysis, hydrophilicity [4] and the commercialization of TiO_2–based products which highlights several points for the future development of TiO_2 photocatalysis [5]. Blake produced a comprehensive bibliography of published work on the heterogeneous photocatalytic removal of organic or inorganic compounds in air and water [27]. In another review article, Walsh *et al.* [28] described the methods of preparation, possible crystal structures and mechanisms of formation of TiO_2 and titanates nanotubes. Grimes *et al.* [29] reviewed the fabrication, properties and solar energy applications of highly ordered TiO_2 nanotube arrays made by anodic oxidation of titanium in fluoride–based electrolytes. Diebold [30] wrote an overview on surface science of TiO_2 with a brief discussion on its bulk structure and bulk defects. In this review the growth of different metals as well as metal oxides on TiO_2 were also discussed. Additionally, recent progress in understanding the surface structure of metals in the 'strong–metal support interaction' state was summarized.

In this book we present a detailed review of the synthesis, properties and application of nanostructured titanium dioxide (NS–TiO_2). First we report the structural, X–ray diffraction and photo–induced properties of NS–TiO_2. Two general approaches for the preparation of NS–TiO_2, namely sol–gel and hydrothermal methods, are presented. The fourth section of this report is devoted to discussions on the applications of NS–TiO_2 which include: the design of dye–sensitized solar cells, hydrogen production and storage, design of antibacterial and self–cleaning agents, electrocatalysis, design of rechargeable batteries, nano–cancer prevention strategies, photocatalytic applications of pure NS–TiO_2, production of TiO_2–based nanoclays and design of modified NS–TiO_2.

In the last section, we describe immobilization methods of TiO_2 nanomaterials on different substrates which include: glass, ceramic, stone, cement, zeolites, metallic and metal oxide materials and polymer substrates.

Chapter 2

Properties of Titanium Dioxide and Its Nanoparticles

2.1. Structural and Crystallographic Properties

Titanium dioxide, CI 77891, also known as *titanium (IV) oxide*, CAS No.: 13463–67–7 has a molecular weight of 79.87 g/mol and represents the naturally occurring oxide with chemical formula TiO_2. When used as a pigment, it is called *"Titanium White"* and *"Pigment White 6"*. Titanium dioxide is extracted from a variety of naturally occurring ores that contain ilmenite, rutile, anatase and leucoxene. These ores are mined from deposits throughout the world. However, most of the titanium dioxide pigment in industry is produced from titanium mineral concentrates by the so–called chloride or sulfate process. This results TiO_2 in the form of rutile or anatase. The primary Titanium White particles are typically between 200–300 nm in diameter, although larger aggregates and agglomerates are also formed [3].

Crystals of titanium dioxide can exist in one of three forms: rutile, anatase or brookite (see Table 1). Their unit cells are shown in Figure 1. In this figure black spheres represent oxygen and the grey spheres represent titanium. In their structures, the basic building block consists of a titanium atom surrounded by six oxygen atoms in a distorted octahedral configuration. In all the three TiO_2 structures, the stacking of the octahedra results in three–fold coordinated oxygen atoms.

Table 1. Crystallographic properties of rutile, anatase and brookite [30, 41].

Crystal structure	Density (kg/m³)	System	Space group	Cell parameters (nm)		
				a	b	c
Rutile	4240	Tetragonal	$D_{4h}^{14} - P4_2/mmm$	0.4584		0.2953
Anatase	3830	Tetragonal	$D_{4a}^{19} - I4_1/amd$	0.3758		0.9514
Brookite	4170	Rhombohedral	$D_{2h}^{15} - Pbca$	0.9166	0.5436	0.5135

The fundamental structural unit in these three TiO_2 crystals forms from TiO_6 octahedron units and has different modes of arrangement as presented in Figure 2. In the rutile form, TiO_6 octahedra link by sharing an edge along the c–axis to create chains. These chains are then interlinked by sharing corner oxygen atoms to form a three–dimensional framework. Conversely, in anatase the three–dimensional framework is generated by edge–shared bonding among TiO_6 octahedrons. This means that octahedra in anatase share four edges and are arranged in zigzag chains. In brookite, the octahedra share both edges and corners forming an orthorhombic structure [30–33].

The monoclinic form of titanium dioxide is titanium dioxide (B) or $TiO_2(B)$. The idealized structure of $TiO_2(B)$ is shown in Figure 3. The three–dimensional framework of $TiO_2(B)$ consists of four edge sharing TiO_6 octahedral subunits (a=1.218 nm, b=0.374 nm, c=0.653 nm) [30, 34]. $TiO_2(B)$ has an advantage over other titanium dioxide polymorphs. Its structure is relatively open and is characterized by significant voids and continuous channels. Because of these properties, $TiO_2(B)$ based nanotubes and nanowires demonstrate great performance in rechargeable lithium batteries [35–38]. High photocatalytic activity was also observed by using TiO_2 nanostructure with polycrystalline phase containing anatase and $TiO_2(B)$ [39]. Although some properties of hydrothermally synthesized $TiO_2(B)$ nanomaterials have been reported [35–40], further studies are required to place them into actual applications.

X–ray Diffraction (XRD) technique is implemented to determine crystal structure as well as crystal grain size of anatase, rutile and brookite. Anatase peaks in X–ray diffraction are occurred at θ=12.65°, 18.9° and 24.054°, the rutile peaks are found at θ=13.75°, 18.1° and 27.2° while brookite peaks are encountered at θ=12.65°, 12.85°, 15.4° and 18.1°. θ represents the X–ray diffraction angle [41–44].

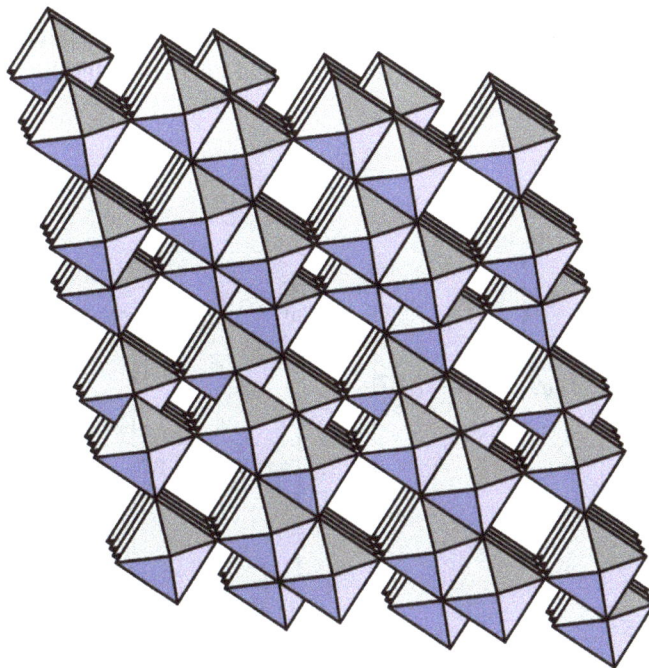

Figure 3. The idealized structure of TiO_2 *(B)*.

2.2. Photocatalytic Properties of *NS–TiO₂*

One of the important properties of the inorganic solid *NS–TiO₂* is its photocatalytic activity. In addition to TiO_2 [45, 46], there is a wide range of metal oxides and sulfides that have been successfully tested in photocatalytic reactions. Among these are ZnO [47], WO_3 [48], WS_2 [49], Fe_2O_3 [50], V_2O_5 [51], CeO_2 [52], CdS [53] and ZnS [54]. Positions and width of energy bands of some of these semiconductors are presented in Figure 4 and compared to those of TiO_2. Interaction of these semiconductors with photons that possess energy equal or higher than the band gap may cause separation of conduction and valence bands as illustrated in Figure 5. This eVent is known as electron–hole pair generation. For TiO_2, this energy can be supplied by photons with energy in the near ultraviolet range. This property promotes TiO_2 as a promising candidate in photocatalysis where solar light can be used as the energy source [55]. Some of the beneficial characteristics of NS–TiO$_2$ include high photocatalytic efficiency, physical and chemical stability, low cost and low toxicity.

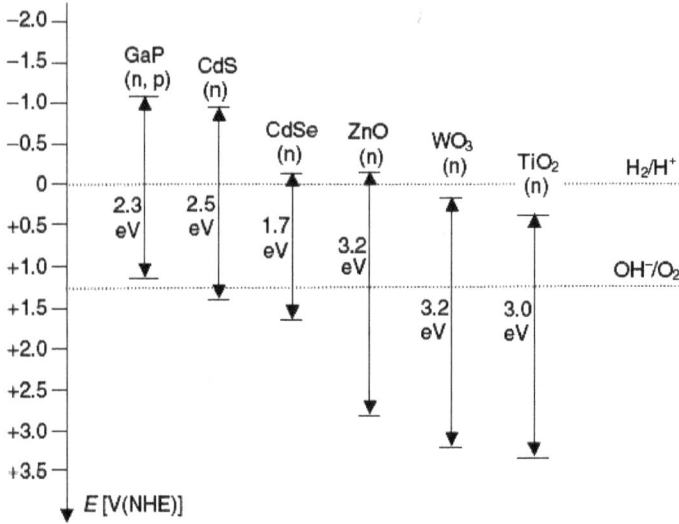

Figure 4. Position and width of energy band of TiO_2 and several other illuminated semiconductors with respect to the electrochemical scale (NHE: normal hydrogen electrode).

As can be observed in Figure 5, when TiO_2 is illuminated with $\lambda <$ 390 nm light, an electron excites out of its energy level and consequently leaves a hole in the valence band. As electrons are promoted from the valence band to the conduction band, they generate electron–hole pairs (Eq. 1) [46, 56]:

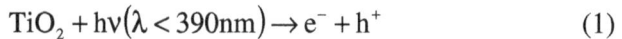

$$TiO_2 + h\nu(\lambda < 390nm) \rightarrow e^- + h^+ \qquad (1)$$

Valence band (h^+) potential is positive enough to generate hydroxyl radicals ($^\bullet$OH) at TiO_2 surface and the conduction band (e^-) potential is negative enough to reduce molecular oxygen as described in the following equations:

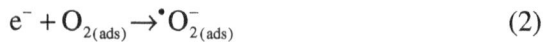

$$e^- + O_{2(ads)} \rightarrow {}^\bullet O_{2(ads)}^- \qquad (2)$$

$$e^- + H^+_{(ads)} \rightarrow {}^\bullet H_{(ads)} \qquad (3)$$

$$h^+ + OH^-_{(ads)} \rightarrow {}^\bullet OH_{(ads)} \qquad \text{(in alkaline solutions)} \qquad (4)$$

$$h^+_{VB} + H_2O_{(ads)} \rightarrow H^+ + {}^\bullet OH_{(ads)} \text{ (in neutral solutions)} \qquad (5)$$

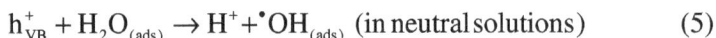

The hydroxyl radical is a powerful oxidizing agent which may attack the organic matters (OM) present at or near the surface of TiO_2. It is capable to degrade toxic and bioresistant compounds into harmless species (e.g. CO_2, H_2O, etc). This decomposition can be explained through the following reactions [56, 57]:

$$h^+_{VB} + OM \rightarrow OM^{\bullet +} \rightarrow \text{Oxidation of OM} \qquad (6)$$

$$^\bullet OH_{(ads)} + OM \rightarrow \text{Degradation of OM} \qquad (7)$$

In addition to the wide energy band gap, TiO_2 exhibits many other interesting properties such as transparency to visible light, high refractive index and a low absorption coefficient. Anatase and rutile, the two principal polymorphs of TiO_2, are associated with energy band gap of 3.2 and 3.1 eV, respectively. It has been pointed out that the photodegradation rate is much more rapid in anatase than in the rutile [58, 59]. This reaction rate is mainly affected by the crystalline state and textural properties such as surface area and particle size. However, these factors often conflict, since a high degree of crystallinity is generally achieved through a high–temperature thermal treatment leading to a reduction in the surface area. Thus, optimal conditions for the synthesis of *NS–TiO₂* have been resulted of the materials of high photoactivity. Since photocatalytic reactions are generally studied in aqueous suspensions, problems arise from the formation of hard agglomerates through the diffusion of reactants and products as well as light absorption. The crystal structure of TiO_2 greatly affects its photocatalytic activity. Amorphous TiO_2 seldom displays photocatalytic activity due to the presence of nonbridging oxygen atoms in the bulk TiO_2. The *Ti–O* atomic arrangement defects could act as recombination centers of photogenerated electron–hole pairs [55].

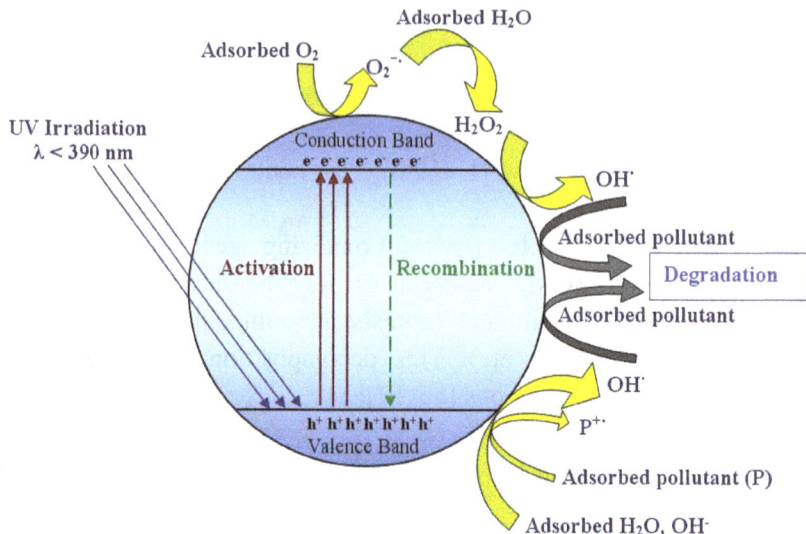

Figure 5. Generation of photocatalytic active species at the surface of TiO_2 nanoparticles. (Adapted from Khataee and Kasiri [1] with permission from publisher, Elsevier. License Number: 2627060102098).

The photocatalytic performance of TiO_2 depends not only on its bulk energy band structure but, to a large extent, on its surface properties. The high photocatalytic activity can be obtained using the photocatalyst with high surface area per mass. Decomposition of methylene blue over TiO_2 photocatalyst films indicates that the photocatalytic activity is strongly dependent on the film surface area of the photocatalyst. These films may have anatase crystal structure with different thickness and surface area. They are prepared through low–pressure metal–organic chemical vapor deposition (LPMOCVD) [60].

The type and density of surface states of NS–TiO_2 are affected by the synthesis process. For instance, a soft mechanical treatment of TiO_2 nanopowder was found to reduce its photocatalytic activity in the reduction of Cr(VI) [61]. On the other hand, treatment in either H_2 or N_2 plasma was found to enhance the activity within the visible–light range for certain reactions [62]. The interplay between processing conditions and photocatalytic activity remains largely a state–of–the–art and is beyond prediction at this point. TiO_2 has typically been calcined or crystallized in oxidizing atmospheres such as air and oxygen. The effects of the inert atmospheres such as N_2, Ar and vacuum (~5 $\times 10^3$ torr), have been overlooked. The calcination atmosphere has been found to have significant effects on the photocatalytic activity of TiO_2 [63]. Calcination in hydrogen or in a vacuum results in a high density of defects and low surface hydroxyl coverage yielding low activity. Calcination in *Ar*, in contrast, enhances visible–light excitation and high hydroxyl coverage leading to higher activity [63].

NS–TiO₂ is successfully used for the photocatalytic remediation of a variety of organic pollutants such as hydrocarbons and chlorinated hydrocarbons (e. g. CCl_4, $CHCl_3$, C_2HCl_3, phenols, chlorinated phenols, surfactants, pesticides, dyes) as well as reductive removal of heavy metals such as Pt^{4+}, Pd^{2+}, Au^{3+}, Rh^{3+} and Cr^{3+} from aqueous solutions. *NS–TiO₂* has also been affective in the destruction of biological organisms such as bacteria, viruses and molds [26, 64–69].

Chapter 3

Preparation of *NS–TiO₂* and Nano–Titanates

In the framework of the rapid development of nanoscience and nanotechnology, the domain of nanostructured materials, such as NS–TiO_2 and nano–titanates requires more academic and industrial research and development studies. Synthesis methods are a major prerequisite to be achieved in this fast developing field [25, 70].

There are several methods to produce NS–TiO_2 and nano–titanates, among this include: (1) hydrothermal method [71, 72]; (2) sol–gel technique [73, 74]; (3) chemical vapor deposition (CVD) [75–77] and physical vapor deposition (PVD) [78, 79]; (4) solvothermal [80, 81]; (5) electrochemical approaches (*e.g.* anodizing of *Ti*) [82–84]; (6) solution combustion [85–87]; (7) microemulsion technique [88, 89]; (8) micelle and inverse micelle methods [90, 91]; (9) combustion flame–chemical vapor condensation process [92, 93]; (10) sonochemical reactions [94] and (11) plasma evaporation [95–97]. Among these manufacturing processes, the most successful are sol–gel and hydrothermal. The advantage of these methods relies on their ability to control the morphology, particle size and crystallinity of the products [71–74]. In the following sections, we describe the production methods of NS–TiO_2, in particular, we emphasize on the sol–gel and hydrothermal techniques.

3.1. Vapor Deposition Method

Recently, vapor deposition methods have been widely explored to fabricate various nanomaterials including NS–TiO_2. In a typical CVD process, thick crystalline TiO_2 films with grain size below 30 nm as well as TiO_2 nanoparticles with size smaller than 10 nm have been prepared through pyrolysis of titanium isopropoxide (TTIP) in a mixed helium/oxygen atmosphere and implementing liquid precursor delivery. When deposited on the cold areas of the reactor at temperatures below 90°C with plasma enhanced CVD, amorphous TiO_2 nanoparticles have been obtained and crystallized with a relatively high surface area to volume ration. This takes place once nanoparticles have been annealed at high temperatures. The disadvantages of this method are its high temperature of the process (~1000°C), significant dimensional changes and geometrical distortions of the products [22, 75–77].

3. 2. Solvothermal Method

This method is almost identical to the hydrothermal process (which is discussed in detail in a later section) except that the solvent used here is nonaqueous. However, the temperature can be elevated much higher than that in the hydrothermal method, since a variety of organic solvents with high boiling points can be selected [22, 80, 81].

3. 3. Electrochemical Approaches

Electrochemical method is commonly employed to produce a coating, usually metallic, on a surface through reduction at the cathode. Anodic oxidation of titanium in various electrolytes, has received significant attention. The effect of synthesis parameters such as current density, electrolyte concentration, applied voltage and the time of anodic oxidation has been extensively studied [82–84]. Among the various groups working on the anodic oxidation process, Grimes and their co–workers [29, 83] have observed the formation of an array of titania nanotubes on a thin titanium foil after an anodization treatment in HF containing aqueous solutions of different concentrations. Constant length arrays of nanotubes with various diameters (25–65 nm) have been produced under variable anodizing voltages. This group also found that as the voltage was increased; particulate or nodular structures, discrete–hollow cylindrical tubes and sponge like porous structure were observed.

3.4. Solution Combustion Method

Solution combustion method is a single step process that produces nanoparticles characterized by their high surface area. The *NS–TiO₂* produced by this method has been successfully applied in the photodegradation of textile effluents under UV and solar radiation and found to degrade the effluents faster than commercial Degussa P25 catalyst [85]. The higher activity is attributed to the higher hydroxyl ion content on the catalyst surface, crystallinity extended till the surface and reduced energy band gap. The synthesis of $NS–TiO_2$ is completed in a single step with no downstream processing. The TiO_2 obtained by this method has a particle size range of 8–12 nm and a surface area equal to 240 m^2/g. Unlike the classical methods of preparation, the maximum temperature reached in the process is 800°C for a short time making the material crystalline [86]. Because of the short time exposure to high temperature, size growth of TiO_2 is hindered and the phase transition to other phases such as rutile and brookite does not occur [7,8]. In a typical combustion synthesis, the precursor of the catalyst is smoldered with a fuel in solution. The precursor is titanyl nitrate. This is obtained by the nitration of the titanyl hydroxide which in turn is a product of the hydrolysis of titanyl isopropoxide and the fuel is glycine. The stoichiometric amount of fuel and precursor for the complete combustion of titanyl nitrate–glycine redox mixture is dissolved in minimum amount of water. The homogeneous solution of this mixture is combusted in a muffle furnace at 350°C. The combustion process involves dehydration followed by smoldering type combustion. High temperature has been experienced only for a very short period of time minimizing the formation of other phases of titania, thus allowing the formation of pure anatase. This method also involves the liberation of large volumes of gases, nearly seven times the moles of the catalyst, resulting in an increased porosity and higher surface area of the material [85–87].

3.5. Microemulsion Technique

Microemulsion technique is a novel method to prepare ultrafine particles. It has the ability to control the size of particles formed and prevent their aggregation. In a typical study, a water–in–oil (W/O) microemulsion known as Winsor Type II microemulsion has been selected for $NS–TiO_2$ preparation. This technique can provide nanosized particles that are much smaller than an oil–in–water (O/W) or Winsor Type I microemulsion. The procedure to prepare the ultrafine particles by this

W/O microemulsion technique starts with two identical W/O microemulsions. One system dissolves reactant No. 1 whereas the other one dissolves reactant No. 2 in the aqueous cores. Upon mixing these two microemulsion systems, both reactants react with each other as a result of the collision and coalescence of the tiny liquid droplets (reverse micelles) suspended in the microemulsion. Nanoparticles are then produced in the aqueous cores [88, 89].

3.6. Micelle and Inverse Micelle Methods

Recently, surfactant self–assemblies have been employed as soft–templates to control the size and shape of nanoparticles. This is referred to as the 'wet' chemical method. Surfactant molecules can self–assemble to form ordered–structures in solution because of their hydrophilic and lipophilic properties. Using surfactant micelles and microemulsions as nanoreactors has become a common shape–controlled methodology. Micelles containing different surfactants have been used to prepare nanoparticles with the following morphologies: circular, hexagonal, triangle, dishlike, nanowire, rod and sphere [90, 91]. Sui *et al.* [91] reported the self–organization of sea urchin–like polyaniline/titanium dioxide (PANI)/TiO_2 nanoparticles in Triton–X100 (OP)/hexamethylene/ water reverse micelle. First 100 µl aniline/TiO_2/HCl solution is added to 0.1M OP/hexamethylene/water solution droplet (10 mL) under vigorous stirring to form the OP/hexamethylene/water reverse micelle. Then 100 µL APS/HCl solution ([APS]=0.2 M, [HCl]=1 M) is added to the reverse micelle. The molar ratio of [aniline]:[APS]:[TiO_2] is kept at 2:2:5. Dark green PANI/TiO_2 nanocomposites are obtained in 2 h. The mixture is kept under vigorous stirring at room temperature for 24 h.

Reverse micelle systems (or water–in–oil microemulsions) are used as microreactors to synthesize ultrafine particles with a narrow particle size distribution by controlling the growth process [90]. Reverse micelles are nanometer–scale surfactant associated in colloid shaped structures formed in a nonpolar organic solvent. Polar solvents such as water are easily soluble inside reverse micelle because the inside of the reverse micelles is quite hydrophilic. Reverse micelle systems are thermodynamically stable, isotropic, transparent mixtures of oil and water separated by a thin surfactant monolayer. These systems provide a micro–heterogeneous medium for the generation of nanoparticles. Dimeric micelles are generated in reverse micelle systems due to collision. Dimeric micelles have a short life span. Upon collision a chemical reaction occurs and substance is exchanged. As collisions repeat, further chemical reactions proceed and nucleus generation takes place. As the hydrolyzed species continue colliding, the nucleus grows up to a fine particle.

3.7. Combustion Flame–chemical Vapor Condensation Process

The Combustion Flame–Chemical Vapor Condensation (CF–CVC) process has been developed to produce nanostructure ceramic powders such as $NS–TiO_2$ which can not be easily produced by the Inert Gas Condensation (IGC) process because of its high melting temperatures and/or low vapor pressures. This process involves the pyrolysis of metalorganic precursors instead of evaporation of a solid metal source as in the IGC process. CF–CVC can synthesize nanostructure powders with no agglomeration, uniform particle size distribution and high purity [92, 93]. Kim *et al.* [92] synthesized $NS–TiO_2$ by using titanium(IV) ethoxide, $(Ti(OC_2H_5)_4)$, as the starting metalorganic precursor. The CF–CVC process consists of a reaction chamber and precursor delivery system. The reaction chamber is maintained at a dynamic He gas pressure of 20 torr by high speed pumping. The pyrolysis of metalorganic precursor/carrier gas stream is performed using a flat flame combustor. The temperature of the metalorganic precursor and carrier gas flow rate is maintained at 130°C and 500 cm^3/min, respectively. The combustion heat sources, H_2 and O_2 gases, are fed into the combustor at a flow rate of 2100 cm^3/min and 2700 cm^3/min, respectively. The synthesized $NS–TiO_2$ consists mostly of anatase with a small amount of rutile. The resulting particle size for anatase and rutile is about 20 nm and 60–70 nm, respectively [92].

3.8. Sonochemical Reactions

Sonochemistry, in which powerful ultrasound is used to stimulate chemical processes in liquids, is currently the focus in chemistry, materials science and technology. Sonication of chemical solutions induces novel chemical reactions and physical changes in the aqueous solutions otherwise. The powerful ability of ultrasound to affect chemical changes arises from cavitation phenomenon involving the formation, growth and collapse of bubbles in the liquid. The implosive collapse of bubbles generates localized hot spots through adiabatic compression within the gas phase of the collapsing bubble. Guo *et al.* [94] described a simple method for the direct synthesis of *NS–TiO₂* employing ultrasound irradiation for a short period of time at a low temperature. In a typical synthesis, deionized water is mixed with ethanol and dispersed in a sonication cell. A mixture of tetraisopropyl titanate (TPT) and ethanol is then added into the cell dropwise. The sonication is carried out employing a direct immersion titanium horn in the sonication cell under ambient atmosphere. During the sonication process, the temperature of the reaction mixture (slurry) rises to approximately 90°C. The reaction continues for 3 h to complete the crystallization of *TiO₂*.

3.9. Plasma Evaporation

In plasma evaporation technique a solution of metal salts is atomized into tiny droplets in several microns. The droplets are introduced into a radio frequency generated at atmospheric pressure. The later is referred to as inductively coupled plasma (ICP). At this stage droplets of the solution are partially or totally evaporated, decomposed and ionized to their elements. The high cooling rate (e.g. 5000 K to substrate temperature of 1000 K) causes a limited growth of nuclei from a supersaturation of evaporated materials. Nanometer–scaled clusters are formed in the tail flame of the plasma. These are then deposited and rearranged on substrate at a temperature lower than that of the plasma. This method has some remarkable advantages among which are a high deposition rate, single precursor and a wide choice of source materials. Preparation of anatase and rutile *NS–TiO₂* thin films by this method has been reported [95–97].

3.10. Hydrothermal Processing

This is an efficient and economical method to obtain nanocrystalline inorganic materials [70, 98]. Hydrothermal processing uses the solubility in water of almost all inorganic substances at elevated temperatures and pressures to induce crystallization of the dissolved material from the fluid. As implied in the name of this method, water at elevated temperatures plays an essential role in the precursor material transformation because the vapor pressure of the reactor is much higher and the structure of water at elevated temperatures is different from that at room temperature. The properties of the reactants including their solubility in water and their specific reactivity also change at high temperatures. These characteristics provide more versatility to control the quality of the nanostructured materials, which are not possible at low temperatures. During the production of nanocrystals, parameters such as water pressure, temperature, processing time and the respective precursor–product system can be tuned to maintain a high nucleation rate and an appropriate size distribution [98–100].

The hydrothermal processing method is one of the important techniques to prepare TiO_2 particles of desired size, shape, homogeneity in composition and a high degree of crystallinity at a relatively low processing temperature. The important features are that it favors a decrease in agglomeration between particles, narrow particle size distributions (monodispersed particles), phase homogeneity and controlled particle morphology. The method also provides uniform composition, high purity of the product and control over the shape and size of the particles [101–103].

The production of $NS–TiO_2$ by the hydrothermal method is usually carried out in small autoclaves of Morey type [101–103], provided with Teflon® liners. The operating conditions selected for the production of $NS–TiO_2$ particles are at a temperature less than, or equal to 200°C and at a pressure less than 100 bars. Such pressure–temperature conditions only require the use of autoclaves of simple design.

Several authors have studied the mild hydrothermal production of $NS–TiO_2$ and the influence of various parameters such as temperature, process duration, pressure and pH on the resultant product [101–129]. We have summarized experimental the conditions and morphological properties of $NS–TiO_2$ and other nano–titanates produced by hydrothermal method in Table 2. In hydrothermal processes, in which the experimental temperature was kept at ~150°C, TiO_2 particles with a high degree of crystallinity and the different size and shape (e.g. Nanoparticles, Nanotubes, Nanosheets and Nanofibers) were achieved through a systematic understanding of the hydrothermal chemistry of the media. It is important to state that the size of titania particles is a critical factor for the performance of the material in photocatalytic activity, in which monodispersed nanoparticles are the most suitable. Experimental data demonstrates that particle size is key in the dynamics of the electron/hole (e^-/h^+) recombination process, which offsets the benefits from the ultrahigh surface area of nanocrystalline *TiO$_2$* [101–105].

Several investigators have used the hydrothermal method to engineered prepare *TiO$_2$* nanoparticles [101–103]. Justin *et al.* [101] prepared nanocrystalline titania with particle size of 20–50 nm and specific surface area of 20–80 of m^2/g by hydrothermal treatment of aqueous TiOSO$_4$, H$_2$TiO(C$_2$O$_4$)$_2$ and TiO(NO$_3$)$_2$ solutions (see Table 2). The group studied the photocatalytic behavior of the synthesized *TiO$_2$* nanoparticles in the photodegradation of phenol in water and explored optimal characteristics of this nanomaterial. Justin and co–workers observed that the best photocatalytic activity was encountered by a mixture of rutile (15%) and anatase (85%) sample, prepared by high–temperature hydrolysis of aqueous TiOSO$_4$ solution.

Chae *et al.* [103] reported the preparation of *TiO$_2$* nanoparticles by hydrothermal reaction of titanium alkoxide in an acidic ethanol–water solution. Titanium isopropoxide was added dropwise to a mixture of ethanol, water and nitric acid at a pH=0.7 and reacted at 240 °C for 4 h. The *TiO$_2$* nanoparticles synthesized under this acidic ethanol–water environment were mainly anatase. By adjusting the concentration of Ti precursor and the composition of the solvent system, the size of the particles was controlled to be in the range of 7–25 nm. The photocatalytic efficiency of *TiO$_2$* films prepared from the 7–nm–sized nanoparticles was 1.6 times that of the films derived from Degussa P25 in decomposing gaseous 2–propanol.

Table 2. Experimental conditions and morphological properties of *NS–TiO₂* and titanates produced by hydrothermal method.

Precursor	Crystal size (nm)	Specific surface area (m^2/g)	Crystallographic phase	Morphology	Hydrothermal Temp. (°C)	Time (h)	Ref.
$TiOSO_4$, $H_2TiO(C2O4)_2$ and $TiO(NO_3)_2$	~20–50	~20–80	Rutile (15%) and Anatase (85%)	Nanocrystalline titania powders	150 or 250	10 min to 6 h	[101]
$TiCl_4$	16–42	–	Rutile	Nanoparticles	220	18	[102]
TTIP[a]	7–25	75–190	Anatase	Nanoparticles	240	4	[103]
$TiCl_4$	5–15	2.63	Rutile	Rods, spindles and spherical nanocrystals	40	16 h	[104]
TBO[b]	8	215	Anatase	Mesoporous TiO_2	130	12	[105]
TBO	3–4	193	Anatase	Mesoporous TiO_2	130	12	[106]
TBOT[c]	7	203.8	Anatase	Mesoporous TiO_2 nanofibers	180	10	[107]
Commercial TiO_2 powders (Degussa P25)	Inner diameter 2–6, outer diameter 5–10, length up to 600	–	$H_2Ti_4O_9.H_2O$	Titanate Nanotubes	110	4	[110]
TTIP	Inner diameter ~5, outer diameter ~8, length of 500–700	93	Anatase	TiO_2 Nanotubes (TiNT)	130	20	[111]

[a] TTIP=Titanium (IV) tetraisopropoxide ($Ti(OCH (CH_3)_2)_4$) or ($Ti(OPr)_4$)
[b] TBO: Titanium(IV) butoxide ($Ti(OCH_2CH_2CH_2CH_3)_4$)
[c] TBOT= Tetrabutyl orthotitanate ($Ti(OC_4H_9)_4$)

Table 2 (continued).

Precursor	Crystal size (nm)	Specific surface area (m^2/g)	Crystallographic phase	Morphology	Hydrothermal Temp. (°C)	Time (h)	Ref.
Rutile TiO$_2$	Inner shell diameter ~8, shell spacing ~0.75, Ave. tube diameter ~12.0	154	H$_2$Ti$_3$O$_7$	TitanateNanotubes (tubes are open ended)	140	96	[112]
TTIP	60×800 and 300×900	–	Anatase	Single rod–like, Besom–like, Nanotubes	160	48	[113]
TTIP	100×300	–	H$_2$Ti$_2$O$_5$·?H$_2$O	Nanosheets	160	48	[113]
TBO	~50–100 width, several nms thickness, Ave. pore diameter 3–4	642	Anatase	Nanosheets	130	12	[122]
Anatase TiO$_2$	10000		Anatase	Nanofibres	160	24	[126]
Natural rutile sand	20–50	45	H$_2$Ti$_x$O$_{2x+1}$ (H$_2$Ti$_3$O$_7$)	As–synthesized Nanofibres	150	72	[128]
Natural rutile sand	20–70	20	TiO$_2$ (B)	Nanofibres	150 (Calcined at 400 °C, 4 h)	72	[128]
Natural rutile sand	20–100	10	Anatase	Nanofibres	150 (Calcined at 700 °C, 4 h)	72	[128]
Natural rutile sand	200–1000	2	Rutile	Submicron rod–like	150 (Calcined at 1000 °C, 4 h)	72	[128]

Manorama *et al.* [104] reported a simple and efficient methodology for the low–temperature synthesis of phase–pure nanocrystalline rutile TiO_2 with a crystallite size range of 5–15 nm. The morphology of the nanoparticles was achieved by a simple variation in the hydrothermal process. This variation consisted in using titanium–tetrachloride without mineralizers, additives or templating agents. They evaluated the photocatalytic activity of the synthesized nanocrystals by photodegrading of methyl orange (MO). The group also investigated the morphology and particle size of the synthesized nanocrsytalline rutile TiO_2 by transmission electron microscopy (TEM) as a function of both hydrothermal reaction temperature and time. According to their results, shown in Figure 6, the effect of reaction time (4, 8, 16, 24, 32, 48 h) at a fixed temperature (100°C) is visible in the slow transformation of the rutile nanocrystals from rods to spindles and then to spherical nanocrystals. After 4 h of reaction time, the particles in the sample resulted in a rod like geometry of about 1000 nm in length and 36–72 nm wide, with an average aspect ratio of 15–20. After 8 h of reaction time, the longer rods were seen to transform into shorter ones (aspect ratio 12) with more spindle like characteristics. At 16 h of reaction time, the nanostructures could be clearly identified as still smaller spindles having an aspect ratio of 10. Thereafter, it can be concluded that with increasing reaction time (24, 32 and 48 h) the rutile TiO_2 has a significant morphological transition from spindle like aggregates to well–defined spherical nanoparticles with average particle size of 100 nm.

Mesoporous TiO_2 has also been synthesized through the hydrothermal methods [100, 105–109]. Mesoporous materials are those with pores in the range of 2–50 nm in diameter. Yoshikaw *et al.* [105] synthesized nanocrystalline mesoporous TiO_2 using titanium butoxide $(C_{16}H_{36}O_4T_i)$ as the starting material. XRD, SEM and TEM analyses revealed that the synthesized TiO_2 had anatase structure with crystalline size of about 8 nm. This titania possessed a narrow pore size distribution with constant pore diameter and a high specific surface area of 215 m^2/g (see Table 2). The photocatalytic activity of synthesized TiO_2 was evaluated with photocatalytic H_2 production from water–splitting reaction. Yoshikaw and his group found that the photocatalytic activity of TiO_2 treated with appropriate calcination temperature was considerably higher than that of commercial TiO_2 (Ishihara ST–01). The results indicated that the utilization of mesoporous TiO_2 characterized by the anatase phase, promoted H_2 production.

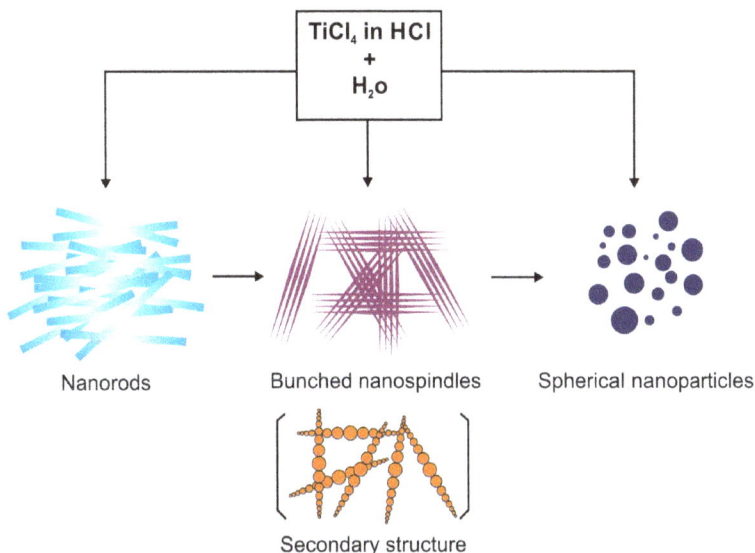

Figure 6. Pictorial representation of the phase–pure rutile titania nanocrystals showing time–temperature dependent morphological transformations under controlled hydrothermal synthesis. (Redrawn from Manorama *et al.* [104] with permission from publisher, Elsevier. License Number: 2627071184341).

In another research report by Yoshikaw and his group, mesoporous anatase TiO_2 nanopowder was synthesized at 130°C for 12 h (see Table 2) [106]. Titanium (IV) butoxide ($C_{16}H_{36}O_4T_i$) was mixed in a 1:1 molar ratio with acetylacetone ($CH_3–CO–CH_2–CO–CH_3$) to slowdown the hydrolysis and the condensation reactions. 40 mL of distilled water was added in the solution and stirred at room temperature for 5 min. The solution was put into a Teflon–lined stainless steel autoclave while stirring and heated at 130°C for 12 h. The final product was naturally cooled to room temperature and washed with 2–propanol and distilled water. This was then dried followed by drying at 100°C for 12 h. The synthesized sample had a narrow pore size distribution with an average pore diameter of about 3–4 nm. The specific surface area of the sample was about 193 m^2/g. Mesoporous anatase TiO_2 nanopowders showed higher photocatalytic activity than the nanorods, nanofibers and commercial TiO_2 nanoparticles.

In another study, bimodal nanocrystalline mesoporous TiO_2 powders with high photocatalytic activity were prepared tetrabutylorthotitanate

(Ti(OC$_4$H$_9$)$_4$, TBOT) as precursor under the optimal hydrothermal conditions (180°C for 10 h) [107]. The photocatalytic activity of the as–prepared TiO_2 powders was evaluated by degradation of acetone (CH$_3$COCH$_3$) under UV–light irradiation at room temperature in the air. The photocatalytic activity of the prepared TiO_2 powders under optimal hydrothermal conditions was three times higher than that of Degussa P25.

TiO_2–based nanotubes with a high specific surface area, ion–exchange ability and photocatalytic ability have been considered for extensive applications [22, 28]. Besides mesoporous TiO_2 nanoparticles, TiO_2 nanotubes (*TiNTs*) have also been synthesized with the hydrothermal method by various research groups [110–121]. For example, titanate nanotubes with inner diameters of 2–6 nm, outer diameters of 5–10 nm and lengths of up to 600 nm were fabricated by directly using commercial TiO_2 powders as the precursors via sonication–hydrothermal combination approach [110]. Ma *et al.* [110] studied titanate nanotubes formation processes during sonication treatment under different sonication powers and times. Scanning electron microscopy (SEM) and transmission electron microscopy (TEM) was used to characterize the morphology of the nanostructures. The chemical composition of the titanate nanotubes was determined by X–ray diffraction (XRD) and energy dispersive X–ray spectroscopy (ERS) analyses (see Table 2). The tubular structure of the titanate nanotubes remained at the calcination temperature of 450 °C, but was completely destroyed at a higher temperature of 600°C.

Yang *et al.* [111] reported the synthesis of ultrahigh crystalline TiO_2 nanotubes by hydrogen peroxide treatment of very low crystalline titania nanotubes (TiNT–as prepared). These were prepared with TiO_2 nanoparticles by hydrothermal methods in an aqueous NaOH solution. The group found that the prepared ultrahigh crystalline TiO_2 nanotubes (TiNT–H$_2$O$_2$) showed comparable crystallinity with high crystalline TiO_2 nanoparticles. TiNT– H$_2$O$_2$ was observed to be multiwalled anatase nanotubes with an average outer diameter of ~8 nm and an inner diameter of ~5 nm. The nanotubes grew along the [001] direction to 500–700 nm in length with an interlayer fringe distance of ca. 0.78 nm (see Table 2). In this study, TiO_2 nanoparticles were prepared in a mixture of a TiO_2/SiO$_2$ with a mole ratio of 90:10. The mole ration was obtained by mixing 52 mL of titanium isopropoxide (TTIP, Ti [OCH (CH₃)₂]₄) and 5.2 mL of tetraethyl orthosilicate (TEOS, Si(OC₂H₅)₄), which were then dissolved in 52 mL of ethanol (99.5%). Upon refluxing the mixture at room temperature for 1 h, 52 mL of ethanol and 40.6 g of 4 M aqueous HCl (36%) were added slowly to the first solution and further stirred at room temperature for 1 h. To precipitate the xerogel, the prepared solusion was placed into an incubator at 80°C for 48 h. The xerogel was dried and calcined in the air at 600°C for 3 h, to become highly crystalline TiO_2 nanoparticles in an anatase form and ca. 20 nm size. To synthesize the titania nanotubes (TiNTs), the prepared and pulverized TiO_2 nanoparticle powders (<38 μm) were treated with an 8 M NaOH aqueous solution in an autoclave Teflon lined high–pressure stainless steel vessel at 130°C and ambient pressure for 20 h. After cooling to room temperature, the precipitated powders were filtrated and washed with 0.1 N HCl and distilled water until the filtrate reached a pH < 7. The filtrated and washed powders were dispersed through sonication for 10 min and dried at ambient temperature. The resulting sample consists of titania nanotubes which are denoted as TiNT–as prepared. For the synthesis of ultrahigh crystalline TiO_2 nanotubes, the TiNT–as prepared powders were treated with a 2 wt % hydrogen peroxide solution under refluxing conditions for a period of time of 4 h at 40°C for. The powders were filtrated and washed with distilled water and dried in an oven at 80°C for 6 h. The resulting sample was ultrahigh crystalline titania nanotubes, known as TiNT–H₂O₂. The photocatalytic activity of TiNT–H₂O₂ was 2–fold higher than TiO_2–P25 (Degussa) in the photocatalytic oxidation of trimethylamine gas under UV irradiation.

Yu *et al.* [112] reported the fabrication of nanocomposites of one–dimensional (1D) titanate nanotubes and rutile nanocrystals by hydrothermal treatment of bulky rutile TiO_2 powders. It was performed in a 10M NaOH solution without using any templates and catalysts. The inner and outer diameters of the nanotubes were approximately 5 and 8 nm, respectively (Table 2). TEM and HRTEM images of the products revealed that many rutile nanocrystals of 5–10 nm attached to the outer surface of the titanate nanotubes. Some rutile nanocrystals of about 5 nm also existed in the interior of the nanotubes. This generates an interesting composite structure that possesses both the surface properties of rutile nanocrystals and the morphology and mechanical properties of titanate nanotubes.

Weng *et al.* [113] studied the effect of tetramethylammonium cations (TMA^+) on TiO_2 crystal morphology under hydrothermal conditions. The as–synthesized samples were characterized by XRD, TEM and SEM methods (see Table 2). The observed morphologies include besom–like particle, nanosheet and nanotubes. The mechanism to accelerate the formation of nanotube in the base of NaOH/TMAOH mixture is illustrated in Figure 7. Bulk TiO_2 is first exfoliated to be layered protonic titanate by the mineralization effect of Na^+. In the presence of TMA^+ cations, the separation of layered protonic titanate is accelerated by intercalating TMA^+ cations in layered titanate. As a result of the presence of more layered titanate in the hydrothermal solution, nanotubes are formed ahead of schedule by curliness of layered titanate. Thus, the mechanism through which TMA^+ cations affect crystal growth in the conditions of this study is different.

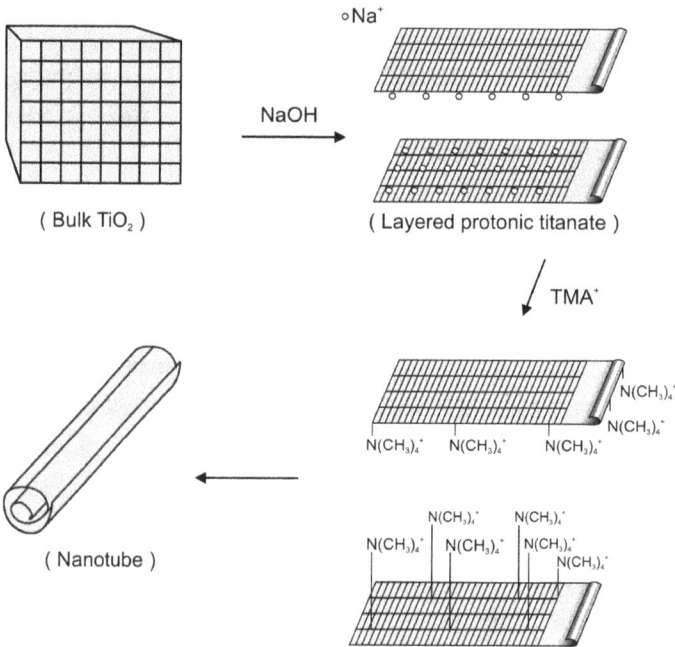

Figure 7. Scheme illustrating the process of nanotube formation in the presence of tetramethylammonium cations (TMA^+) in the alkaline base by hydrothermal synthesis. (Redrawn from Weng *et al.* [113] with permission from publisher, Elsevier. License Number: 2627080246987).

Pavasupree *et al.* [122] reported the preparation of high surface area *TiO₂* nanosheet (e.g. 642 m²/g) through a hydrothermal process at 130°C for 12 h. Briefly, titanium (IV) butoxide (TiBu) was mixed in a similar mole ratio with acetylacetone (ACA) to slowdown the hydrolysis and the condensation reactions. 40 mL of distilled water were added into the solution while stirring at room temperature for 5 min. 30 mL of ammonia in an aqueous form (28%) were added into the solution. This was placed into a Teflon–lined stainless steel autoclave and heated at 130°C for 12 h. The container was naturally cooled to room temperature. The obtained product was washed with HCl aqueous solution, 2–propanal and distilled water several times, followed by drying at 100°C for 12 h. The samples were calcined 4 h at 300–700°C in the air condition (see Table 2). According to the results, at first, amorphous *TiO₂* nanoparticles were generated in the TiBu–ACA solution. Many aggregations of *TiO₂* nanoparticles formed after 28% ammonia aqueous solution was added). The later induces nuclei formation which causes *TiO₂* to grow (Figure 8). The nanosheet *TiO₂* formed under hydrothermal treatment because the nanotubes (from nanosheets rolling technique) can be synthesized at 110–120°C for 48–72 h and diluted base treatment generates thin, curled sheet materials. The nanosheet structure after calcination was destroyed and changed to nanorods/nanoparticles composites at high temperature [122–125].

TiO₂ nanofibers have also been synthesized through hydrothermal methods [109, 126–130]. Li *et al.* [126] results showed that the growth of *TiO₂* nanofibers was sensitive to the concentration of NaOH and the heating temperature. They investigated morphologies of the as–prepared products by transmission electron microscopy. The particle size is about 10 μm. Compared with Ma *et al.* [110], who have prepared titanium oxide nanotubes with the outer diameter of 5–10 nm and a length of 600 nm via sonication–hydrothermal combination approach (e.g., for 4 h at 110°C) with a 10M NaOH aqueous solution, *TiO₂* nanofibers have been prepared with lower concentration of NaOH and higher temperature (see Table 2).

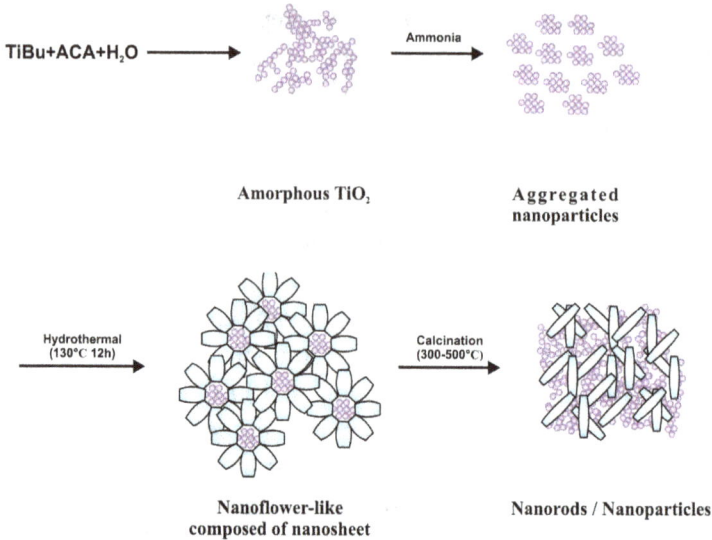

TiBu+ACA+H₂O ⟶ ... Ammonia ⟶ ...

Amorphous TiO₂   Aggregated
                                  nanoparticles

Hydrothermal   Calcination
(130°C 12h) ⟶   (300-500°C) ⟶

Nanoflower-like   Nanorods / Nanoparticles
composed of nanosheet

Figure 8. Schematic representation of the growth diagram of nanosheet and nanorods/nanoparticles of TiO_2. (Redrawn from Pavasupree *et al.* [122] with permission from publisher, Elsevier. License Number: 2627090439895).

Synthesis of titanate, $TiO_2(B)$ and anatase TiO_2 nanofibers from natural rutile sand have been reported by Susuki *et al.* [127, 128]. Titanate nanofibers were synthesized at 150°C for 72 h using 96.0% TiO_2 yielded in Australia, Tiwest Sales Pty. Ltd., Bentley, Australia, as the starting materials [128, 129]. $TiO_2(B)$ and anatase TiO_2 (high crystallinity) nanofibers with diameters of 20–100 nm and the lengths of 10–100 μm were obtained from calcined titanate nanofibers for 4 h at 400 and 700°C (in the air), respectively. At temperatures higher than 900°C, the nanoparticles began to change into rutile–type TiO_2 with a rod–like structure (see Table 2). In a typical synthesis, 300 mg of natural rutile sand were put into a Teflon–lined stainless steel autoclave. To this 50 mL of 10 M NaOH were added and heated at 150°C for 72 h while stirring. Upon cooling to room temperature, the obtained product was washed with HCl solution and distilled water several times followed by freezed drying. The samples were calcined for 4 h at 120–1000°C in the air condition [128].

3.11. Sol–Gel Technology

This method has a relatively long history. It started with processing of oxide materials including glass and ceramics about 30 years ago [130]. Since then, the technology has been employed not only in preparation of oxides, but also in the preparation of non–oxide materials such as nitrides, carbides, fluorides and sulfides as well as oxynitride and oxycarbide glasses. Processing of organic–inorganic materials is now a very active field of research. This has even expanded to the field of biotechnology where it applies in the research of encapsulation of enzymes, antibodies and bacteria. This technology is a versatile tool that makes it possible the production of a wide variety of metal oxide nanostructures with novel properties [130, 131].

Three approaches are employed to make sol–gel monoliths: (1) gelation of a solution of colloidal powders; (2) hydrolysis and polycondensation of alkoxide or nitrate precursors followed by hypercritical drying of gels; and (3) hydrolysis and polycondensation of alkoxide precursors followed by aging and drying under ambient atmosphere. Sols are dispersions of colloidal particles in a liquid. Colloids are solid particles with diameters of 1–100 nm. A gel is an interconnected, rigid network with pores in the submicrometer range and polymeric chains whose average length is greater than a micrometer. The term "gel" embraces a diversity of combinations of substances that can be classified into four categories: (1) well–ordered lamellar structures; (2) covalent polymeric networks, completely disordered; (3) polymer networks formed through physical aggregation, predominantly disordered; and (4) particular disordered structures [15, 130].

Recently, sol–gel processes became a technique for the preparation of *NS–TiO₂*. It has been demonstrated that through sol–gel processes, the physico–chemical and electrochemical properties of *TiO₂* can be modified to improve its efficiency. It provides a simple and easy means of synthesizing nanoparticles at ambient temperature under atmospheric pressure and this technique does not require complicated set–up. Since this method is a solution process, it has all the advantages over other preparation techniques in terms of purity, homogeneity, control of particle size, felicity and flexibility in introducing dopants in large concentrations, stoichiometry control, ease of processing and composition control. Through sol–gel process, the growth of TiO_2 colloids in nanometer range can be effectively controlled by hydrolysis and condensation of titanium alkoxides in an aqueous medium [132–137]. A disadvantage of this method is that in most cases the template material is scarified and needs to be destroyed after synthesis, leading to an increase in the cost of materials [28, 138].

The preparation of $NS–TiO_2$ by sol–gel method usually involves controlled sol–gel hydrolysis of solutions of titanium–containing compounds in the presence of templating agents, followed by polymerization of TiO_2 in the self–assembled template molecules or deposition of TiO_2 onto the surface of the template aggregates. The following stage consists in the selective removal of the templating agent and calcination of the sample. Template syntheses of $NS–TiO_2$ can be separated into several groups according to the type of template molecules used (see Table 3).

As it was mentioned previously, this process normally proceeds via an acid–catalyzed hydrolysis step of a titanium precursor, such as titanium(IV) alkoxide, followed by condensation. The development of Ti–O–Ti chains is favored with low content of water, low hydrolysis rates and excess titanium alkoxide in the reaction mixture. Three dimensional polymeric skeletons with close packing result from the development of Ti–O–Ti chains. The formation of $Ti(OH)_4$ is favored with high hydrolysis rates for a medium amount of water. The presence of a large quantity of Ti–OH and insufficient development of three–dimensional polymeric skeletons lead to loosely packed first–order particles. Polymeric Ti–O–Ti chains are developed in the presence of a large excess of water. Closely packed first order particles are yielded via a three–dimensionally developed gel skeleton [22, 28, 132–135].

Table 3. Experimental conditions and morphological properties of *NS–TiO₂* and titanates produced by sol–gel method.

Precursor	Hydrolyzing agent	Calcination temperature (°C) and time(h)	Crystal size (nm)	Specific surface area (m^2/g)	Crystallographic phase	Remarks	Ref.
TTIP	Methanol	500, 5	17	69	Anatase (69%) and Rutile (31%)	Nanocrystalline TiO₂	[139]
TTIP	Isopropyl alcohol	500, 5	12.6	84	Anatase (74%) and Rutile (26%)	Nanocrystalline TiO₂	[139]
TTIP	Glacial acetic acid	500, 5	8.3	107	Anatase (82%) and Rutile (18%)	Nanocrystalline TiO₂	[139]
Tetra–n–butyl–titanate	Deionized water	600, 3	36	–	Anatase	TiO₂ Nanopowder	[140]
TBO	Ethanol solvent, acetyl–acetone as a complexing and chelating	600, 3	15	–	Anatase	Thin Film of Nanocrystalline TiO₂	[141]
TBO	Ethanol solvent, acetyl–acetone as complexing and chelating	Microwave (2.45 GHz) at 600 W for 10 min	9	–	Anatase	Thin Film of Nanocrystalline TiO₂	[141]
TTIP	H₂O and HNO₃	400, 3	–	–	Anatase	Thin TiO₂ films	[143]
Titanium oxyhydroxide	Electrochemically induced sol–gel process	450, 24	Diameter ~ 40	–	Anatase	Single–crystalline nanowires	[147]
TTIP	Nonhydrolytic sol–gel With oleic acid	270, 2	3.4 diameter, 38 length	198	Anatase	Nanorods	[152]

Table 3 (continued).

Precursor	Hydrolyzing agent	Calcination temperature (°C) and time(h)	Crystal size (nm)	Specific surface area (m²/g)	Crystallographic phase	Remarks	Ref.
TTIP	Oleic acid (C₁₈H₃₃COOH)	500, 0.5	4 diameter, 40 length	–	Anatase	Nanorods	[153]
TTIP	Acetyl acetone, water and ethyl alcohol	400, 10 h		–	Anatase	Nanorods	[156]
TTIP	Acetyl acetone, water and ethyl alcohol	700, 30	100–300 diameter	–	Rutile	Nanorods	[156]
Titanium tetrachloride (TiCl₄)	Water as diluting, Hydrochloric acid as catalyst. Isopropyl alcohol as co–solvent	Without calcination, 83, 15	10	190	Brookite	Nanoparticles	[157]
TTIP	Ethanol/water mixtures	500, 3	–	–	Brookite	Nanoparticles	[159]
TTIP	Distilled Water, TEOA¹	Aged at 140 °C for 3 days in autoclave	Shape control : Ellipsoidal particles (29.1×11.4 nm; 37.0 m²g⁻¹) Cubic particles (15.0 nm; 43.0 m²g⁻¹)		Anatase	Nanoparticles	[165]
TTIP	Distilled water, TEOA	Aged at 140 °C for 72 h in autoclave	controllable: 5–30	–	Anatase	Nanoparticles	[169]

¹ TEOA = Triethanolamine (N(CH₂CH₂OH)₃))

Many groups have used the sol–gel method to prepare *NS–TiO$_2$* [139–150]. For example, Murugesan *et al.* [139] used the typical synthesis procedure for *NS–TiO$_2$* as follows. Titanium (IV) isopropoxide ($C_{12}H_{28}O_4Ti$), glacial acetic acid and water were mixed in a molar ratio of 1:10:350. Titanium (IV) isopropoxide (18.6 mL) was hydrolyzed using 35.8 mL glacial acetic acid at 0°C. To this solution water (395 mL) was added drop wise under vigorous stirring for 1 h. Subsequently the solution was sonicated for 30 min followed by stirring for another 5 h until a clear solution of *TiO$_2$* nanocrystals was formed. The solution was then placed in an oven at 70°C for a period of 12 h to allow an aging process. The gel was then dried at 100°C and subsequently the catalyst was crushed into fine powder and calcined in a muffle furnace at 500°C for 5 h. Through this method, different anatase to rutile ratios of nanocrystalline *TiO$_2$* with high surface area were obtained (see Table 3).

Preparation and characterization of nano–*TiO$_2$* powder using a sol–gel method was also reported by Li *et al.* [140]. Their aim was to investigate the factors that affect the grain size of nano–*TiO$_2$* powder during synthesis. The results indicated that the preparation conditions such as concentration, pH value, calcination time and calcination temperature had much influences upon the properties of nano–*TiO$_2$* powders. It became clear that among these factors the calcination temperature and pH values were more effective compared with the calcination time and concentration. The grain size tends to increase with increasing temperature and the increase in pH value. Different calcination time was found to produce different effects upon the grain size depending upon calcinations temperature. The higher the calcination temperature, the greater the effect of calcination time upon the grain size.

Besides *TiO$_2$* nanoparticles, thin films of *NS–TiO$_2$* have also been synthesized with sol–gel method by various research groups [141–146]. Phani *et al.* [141] reported deposition of thin films of *TiO$_2$* on polished quartz substrates at room temperature as result of sol–gel dip–coating technique followed by two different annealing treatment methods. One set by conventional annealing at 600°C for 3 h and other set exposed to microwave (2.45 GHz) radiation at 600 W for 10 min (see table 3). Microwave processing caused crystallization to occur at a lower temperature. Therefore, microwave processing has the potential to reduce the time, cost and required input energy for the production of $NS-TiO_2$ for dye sensitized solar cells.

Addamo *et al.* [143] prepared thin TiO_2 films with the dip–coating technique and sols derived from titanium tetraisopropoxide. TiO_2 films were formed on glass substrates previously covered by a SiO_2 layer obtained from a tetraethylortosilicate sol. Upon thermally treating the films at 673 K, they consisted of TiO_2 anatase. The photoactivity of the various films has been tested by the photo–oxidation of 2–propanol in gas solid regime. The photoactivity results indicated that the TiO_2 films were efficient in degrading 2–propanol under UV illumination.

Sol–gel preparation of TiO_2 nanowire arrays have been reported by various research groups [147–151]. Miao *et al.* [147] reported the fabrication of highly ordered TiO_2 single–crystalline nanowire arrays within the pores of anodic aluminum oxide template by a cathodically induced sol–gel method. Raman spectra and TEM investigations confirmed that the nanowires were composed of pure anatase TiO_2 with a uniform tetragonal single–crystal structure. In another example, ordered anatase TiO_2 nanowire arrays have been successfully fabricated in the nanochannels of a porous anodic alumina membrane by electrophoretic deposition [148]. The photocatalytic activities of TiO_2 nanowire arrays were characterized by quantifying the degradation of Rhodamine B dye solution. Compared with TiO_2 nanowire arrays and thin film prepared through sol–gel deposition, higher photocatalytic activity was obtained by TiO_2 nanowire arrays by using electrophoretic deposition. This was attributed to larger surface areas and greater energy gap.

Nanorods of TiO_2 have been synthesized by sol–gel method [152–156]. Hyeon *et al.* [152] reported a simple method of synthesizing a large quantity of TiO_2 nanorods. A nonhydrolytic sol–gel reaction between titanium(IV) isopropoxide and oleic acid at 270°C generated 3.4 nm (diameter) ×38 nm (length) sized TiO_2 nanorods. According to the XRD spectra and selected–area electron diffraction patterns combined with high–resolution TEM images, the group found that the TiO_2 nanorods were characterized by a crystalline anatase structure grown along the [001] direction. The group claims that the diameters of the TiO_2 nanorods could be controlled by adding 1–hexadecylamine to the reaction mixture as a co–surfactant. Their results also indicated that TiO_2 nanorods with average sizes of 2.7 nm×28 nm, 2.2 nm×32 nm and 2.0 nm×39 nm were obtained using 1, 5 and 10 mmol of 1–hexadecylamine, respectively. The resulting *TiO$_2$* nanorods exhibited higher photocatalytic activity than a commercial P25 photocatalyst for the photocatalytic inactivation of *E. coli*. This enhanced photocatalytic activity seems to be a result of the nanorods high surface area, large amount of surface hydroxyl groups and the decreased band gap of *TiO$_2$* nanorods.

Hashimoto *et al.* [153] developed a new method to fabricate an array of *TiO$_2$* nanorod assemblies on a flat *TiO$_2$* surface. The *TiO$_2$* nanostructures with a height of approximately 40 nm were synthesized via a low temperature sol–gel reaction in a reversed micelle system. Typically, a total of 40 mL of Oleic acid (C$_{18}$H$_{33}$COOH or OLEA) was placed in a 50 mL two–neck flask and connected to a vacuum to dehydrate at 90°C for 1 h. A total of 1.5 mL of tetraisopropyl orthotitanate was added under a nitrogen atmosphere and stirred at 90°C for 5 min. When the solution turned to pale yellow, 0.75 g of trimethylamine–*N*–oxide dihydrate (($CH_3)_3NO·2H_2O$ or TMAO) was added. After the TMAO dissolved in the solution, 5 mL of the solution was taken into a 15 mL PFA container where the substrate was already set with Indium tin oxide (ITO) side facing up. A total of 0.6 mL of water was injected into the PFA container and allowed to age at 90°C for 12 h. After being aging in the oven, the substrate was carefully washed with hexane several times and then immersed into hexane for several hours. After the substrate was dried, it was heated at 500°C for 30 min to remove the residual organic compounds such as OLEA. Cross–sectional TEM observation revealed that each nanostructure is an assembly of nanorods with a diameter and length of approximately 4 nm and 40 nm, respectively. The fabricated array of *TiO$_2$* nanorods was applied to construct photovoltaic devices combined with a semiconducting polymer.

Besides anatase and rutile nanoparticles, brookite phase *TiO$_2$* nanoparticles have also been synthesized by sol–gel method [157–161]. Lee and Bhave [157] reported the preparation of nanocrystalline brookite titanium dioxide particles by sol process under ambient condition. Titanium tetrachloride was the precursor in water with isopropanol as the co–solvent in hydrochloric acid. The formed gel mass was peptized and crystallized under refluxing condition. The effect of various parameters, such as temperature, time, water to alcohol ratio and pH of the sol on the phase obtained were investigated. The refluxing temperature of 83°C with the refluxing time of 15 h led to the formation of pure brookite phase (see Table 3). According to their results, the refluxing time and temperature play a very vital role in controlling the particle size and the phase of the particles obtained. The increase or decrease in the temperature, above or below 83°C led to phase impurity to the pure brookite nanoparticles. Acidic nature of the sol was necessary for obtaining brookite phase [157].

Nanoparticle size, shape, microstructure as well as phase are known to strongly influence nanocrystal properties such as color, chemical reactivity and catalytic behavior [161–163]. Previous works addressing the link between sol–gel variables (e.g. temperature, pH, addition rates) and particle properties have demonstrated that small changes in synthesis procedures can yield large changes in the physical and chemical properties of the products [164–166]. Sugimoto *et al.* indicated that the morphology of titania particles produced in the presence of triethanolamine was pH dependent. At pH 9.6, cubic particles were produced, whereas at pH 11.5, ellipsoidal particles were produced. Understanding the link between synthesis variables and the physical and chemical properties of the product materials is critical to achieving control over the properties of nanoparticles [165].

Sugimoto *et al.* [165–168] reported the shape control formation of uniform anatase–type TiO_2 nanoparticles by phase transformation of a $Ti(OH)_4$ gel matrix in the presence of shape controllers. Triethanolamine (TEOA) was found to change the morphology of TiO_2 particles from cuboidal to ellipsoidal at pH above 11. Similar shape controls to yield ellipsoidal particles with a high aspect ratio was also achieved with other primary amines, such as ethylenediamine (ED), trimethylenediamine (TMD) and triethylenetetramine (TETA). According to their results, TEOA changes the morphology of anatase TiO_2 particles from cuboidal to ellipsoidal at pH above 11 by specific adsorption to the growing TiO_2 particles. The mechanism of the shape control can be explained in terms of the reduction of the growth rate of the crystal planes parallel to the c–axis of the tetragonal system by the specific adsorption of TEOA to these planes [165].

```
┌─────────────────────────┐
│  Titanium Isopropoxide  │   (TIPO: Ti[OCH(CH₃)₂]₄)
└─────────────────────────┘
             │
             │◄─────────── Triethanolamine (TEOA: N(C₂H₄OH)₃)
             │
             │             [TEOA]/[TIPO]=2
             │
             │                        C₂H₄OH
             │                          |
┌─────────────────┐        N(C₂H₄O)₃Ti-OH₄C₂-N-C₂H₄O-Ti(OH₄C₂)₃N
│ Stable Complex  │
└─────────────────┘
             │
             │◄─────────── H₂O (HClO₄ or NaOH)
             │
             │             First aging ( 100°C , 24h )
             │
┌─────────────────┐
│  Ti(OH)₄ Gel    │        ([Ti(IV)]=0.25 mol dm⁻³)
└─────────────────┘
             │
             │             Second  aging ( 140°C , 72h )
             │
┌─────────────────┐
│  TiO₂ (anatase) │
└─────────────────┘
```

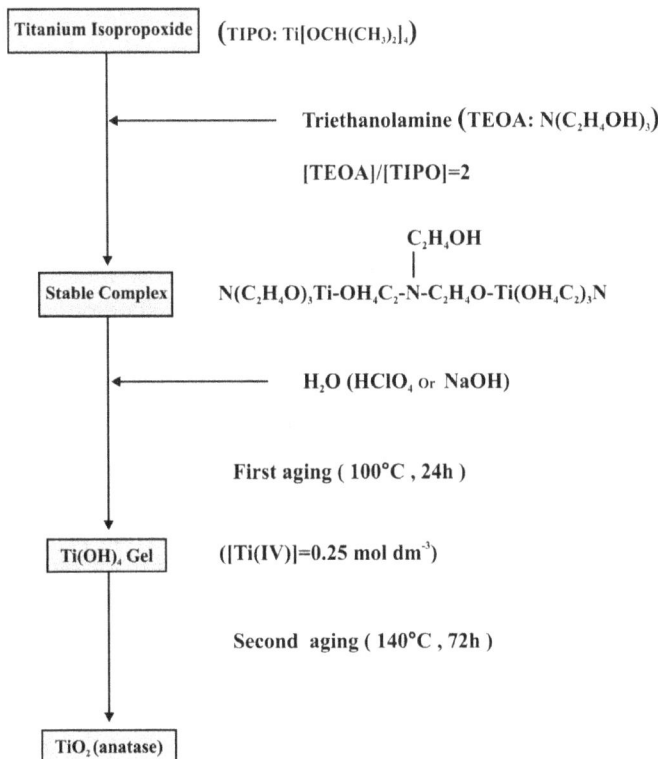

Figure 9. Flow chart for size controlled preparation of uniform anatase–type *TiO$_2$* nanoparticles by the gel–sol process. (Redrawn from Sugimoto *et al.* [169] with permission from publisher, Elsevier. License Number: 2627090902750).

Sugimoto *et al.* [169] also studied the size control preparation of uniform anatase–type *TiO$_2$* nanoparticles by the gel–sol process from a condensed Ti(OH)$_4$ gel preformed by the hydrolysis of a Ti–TEOA complex. The procedure is presented in Figure 9. The particle size of the anatase *TiO$_2$* was increased from ca. 5 to 30 nm with pH increasing from 0.6 to 12 by aging the Ti(OH)$_4$ gel at 140°C for 72 h. TEOA appeared to enhance the pH effect on the nucleation rate of anatase *TiO$_2$* particles by adsorption onto their embryos, leading to a wide range of size control.

Chapter 4

Applications of Nanostructured Titanium Dioxide (*NS–TiO₂*)

NS–TiO$_2$ materials, with a dimension of less than 100 nm, have recently emerged. They include nanoparticles in shapes of spheroids, nanotubes, nanorods, nanowires, nanosheets and nanofibers [24, 25, 170]. Table 4 summarizes potential applications of *NS–TiO$_2$*. The unique physicochemical properties for all these forms of *NS–TiO$_2$*, render this material a promising future in many applications. As indicated in Table 4, *TiO$_2$*, which is the most common compound of titanium, is often used in a wide range of applications including antibacterial purposes, self–cleaning coatings and cancer treatment, solar cells and photocatalysis. We describe each of these applications in this chapter.

Table 4. The possible applications of nanostructured *TiO$_2$* (*NS–TiO$_2$*).

Application	Example	References
Photovoltaic cells	Dye–sensitized solar cells	[185–190]
Hydrogen production	Photocatalytic splitting of water, Hydrogen production from natural seawater, Separate evolution of H_2 and O_2 from water under visible light irradiation	[202, 210–223]
Hydrogen storage	Reversible storage of H_2 on nanotubular TiO_2 arrays	[234–246]
Sensor	Sensing humidity, H_2O_2, VOC, ammonia, Oxygen and Hydrogen, Monitoring chemical oxygen demand (COD) and determination sensor to flue gas on spark ignition engine	[257–302]
Batteries	Lithium–ion battery, Polyaniline/TiO_2 composite in rechargeable battery	[310–324]
Cancer prevention and treatment	Controlled drug–release Reservoir with Temozolomide (TMZ) Photocatalytic Cancer–Cell Treatment	[325–338]

Table 4 (Continued).

Antibacterial and self–cleaning activity	Materials for residential and office buildings, Road construction, vehicles, hospitals, air cleaning and Self cleaning glass, (see Table 5)	[4, 5, 343–350]
Electrocatalysis	Electrocatalysis of methanol, Electrocatalytic oxidation of nitric oxide at TiO_2–Au nanocomposite, Electrochemical wastewater treatment	[351–356]

4.1. Dye–Sensitized Solar Cells

The interest in dye–sensitized nanocrystalline TiO_2 solar cells has grown considerably in recent years from fundamental and applied perspectives [171–175]. Although there are many other semiconducting metal oxides such as ZrO_2 [176], ZnO [177], SnO_2 [178], Nb_2O_5 [179], CeO_2 [180] and $SrTiO_3$ [181], titanium dioxide has been the most widely used product for solar cells. It exhibits efficient light harvesting capacity as well as low production cost. Particularly, intense research has been undertaken on the textural properties of nanocrystalline TiO_2 films since they play a key role in light harvesting of solar cell [182–184].

A typical dye–sensitized solar cell (DSSC) consists of two conductive transparent glasses, ruthenium dye–sensitized TiO_2 film, a platinum catalyst layer and a liquid electrolyte containing I^-/I_3^- redox couple (Figure 10) [185, 186]. Figures 10 and 11 illustrate the structure and operating principles of the dye–sensitized solar cell. At the heart of the system, there is a mesoporous TiO_2 film with a monolayer of the charge transfer dye attached to its surface. The film is placed in contact with a redox electrolyte or an organic hole conductor. Photoexcitation of the dye injects an electron into the conduction band of TiO_2. The electron can be conducted to the external circuit to drive the load and produce electric power. The original state of the dye is subsequently restored by an electron donation from the electrolyte. This is sually an organic solvent containing a redox system such as the iodide/triiodide couple. The regeneration of the sensitizer by iodide prevents the recapture of the conduction band electron by the oxidized dye. The iodide is regenerated in turn by the reduction of triiodide at the counter electrode, with the circuit being completed via electron migration through the external load. The voltage generated under illumination corresponds to the difference between the Fermi level of TiO_2 and the redox potential of the electrolyte. Overall, the device generates electric power from light without suffering any permanent chemical transformation [187, 188].

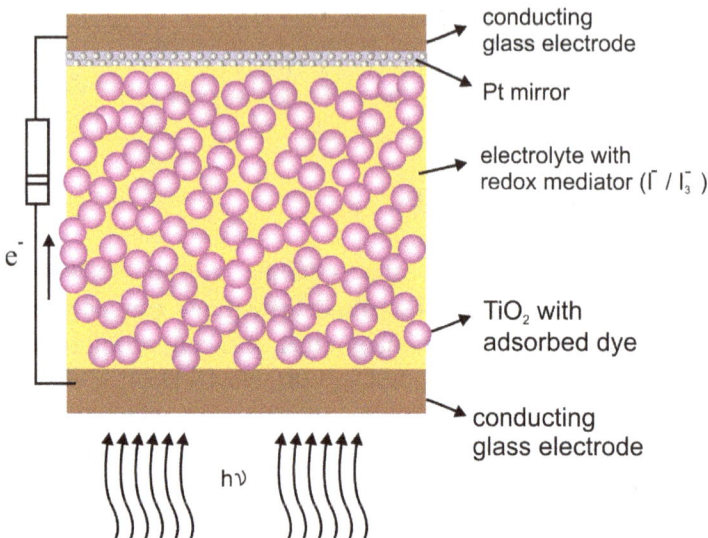

Figure 10. Structure of the dye–sensitized nanocrystalline solar cell.

Grätzel *et al.* [172, 187, 188] explained the basis of the photocurrent and photovoltage in nanocrystalline mesoporous TiO_2 dye–sensitized solar cells. According to their findings, when TiO_2 solar cells are exposed to light three main reactions take place:

(a) electron injection from the dye excited state to conduction band of TiO_2,

(b) hot electron injection or relaxation/cooling processes of the hot electron in the conduction band and in trap states,

(c) recombination between conduction band electron and dye cation and/or capture by redox mediator species [187–189].

The group also found that electron injection onto the TiO_2 conduction band could occur on a fast or ultra fast time scale. The kinetics of electron transfer in the dye–sensitized nanocrystalline TiO_2 solar cell is summarized by following equations. Recombination reaction takes place in a hundredth of a nanosecond. Regeneration of the dye cation by the iodide system is faster than the recombination reaction.

Photovoltaic performances suffer strongly due to critical reactions between conduction band electrons and radical anion $I_2^{\bullet-}$ (i.e. on average of hundred nanosecond time scale in regime I) [172].

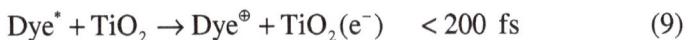

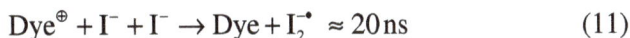

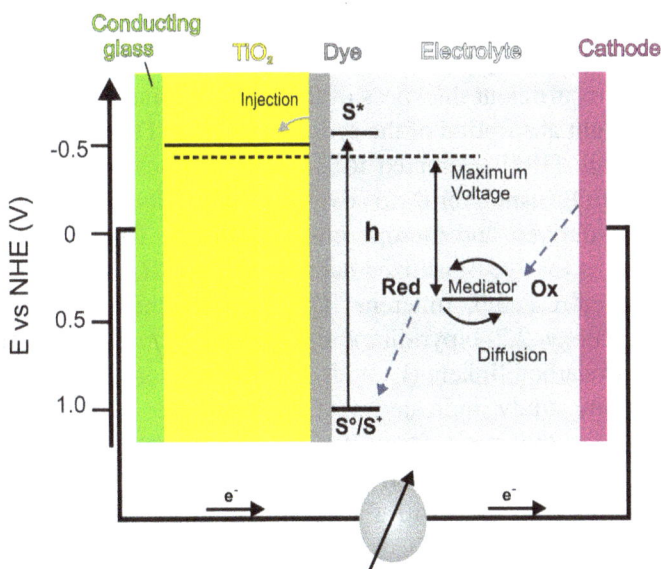

$$Dye^* \rightarrow Dye \quad \approx 30 \text{ ns} \tag{8}$$

$$Dye^* + TiO_2 \rightarrow Dye^{\oplus} + TiO_2(e^-) \quad < 200 \text{ fs} \tag{9}$$

$$Dye^{\oplus} + TiO_2(e^-) \rightarrow Dye + TiO_2 \quad < 500 \text{ ns} \tag{10}$$

$$Dye^{\oplus} + I^- + I^- \rightarrow Dye + I_2^{\bullet} \quad \approx 20 \text{ ns} \tag{11}$$

$$I_2^{\bullet} + e^- \rightarrow 2I^- \quad \approx 100 \text{ ns} \tag{12}$$

$$2I_2^{\bullet} \rightarrow I^- + I_3^- \quad \mu s \tag{13}$$

$$I_3^- + 2e^- \rightarrow 3I^- \quad ms \tag{14}$$

Figure 11. Principle of operation of the dye–sensitized nanocrystalline solar cell. Photo–excitation of the sensitizer (S) is followed by electron injection into the conduction band of an oxide semiconductor film. The dye molecule is regenerated by the redox system, which itself is regenerated at the counter–electrode by electrons passed through the load. Potentials are referred to the normal hydrogen electrode (NHE). The energy levels drawn match the redox potentials of the standard N$_3$ sensitizer ground state and the iodide/triiodide couple. (Redrawn from Grätzel [187] with permission from publisher, Elsevier. License Number: 2627070632803).

Attempts have been made to replace the components with new electrode materials, alternative sensitizers and different types of electrolytes. In most cases studies were aimed at achieving advanced properties and applying DSSC in order to meet the requirements of commercial applications and/or to reduce the production cost. An important issue in cost reduction is the fabrication of DSSC on flexible, transparent, conductive plastic substrates [190–192]. The nanocrystalline TiO_2 film is considered to be the most promising material for the electrode of dye–sensitized solar cells to generate high performance [193–195].

Kallioinen *et al.* [196] studied the dynamic preparation of nanoporous TiO_2 films for the fabrication of DSSCs. The dynamic preparation includes spreading and pressing of the films with roll–to–roll compatible methods. To spread of TiO_2 paste, gravure printing was selected. This technique is used in the packing and printed matter industry and it is a roll–to–roll type preparation method. The results indicated that the efficiency of cells with gravure printed TiO_2 film was quite low. The best light conversion efficiency was only 1.7%. This was the result of insufficient thickness of the gravure printed TiO_2 film which led to poor light absorption of the sensitized film in DSSC.

Choy *et al.* [197] attempted to fabricate a 'house–of–cards (HOC)' structured TiO_2 nanohybrid to overcome the disadvantage of both nanoparticle–derived and mesoporous TiO_2 films. This structure offers salient features for dye–sensitized nanocrystalline TiO_2 solar cell.

In another research, fullerene (C_{60}) was attached to N3 dye (cis–bis(4,4–dicarboxy–2,2–bipyridine)dithiocyanato ruthenium(II)) via diaminohydrocarbon linkers (L) with different carbon chain lengths. The results of this study indicated that in the case of the linker 1,6–diaminohexane, the current density, applied potential and conversion efficiency of the pertaining cell were 11.75 mA/cm^2, 0.70 V and 4.5%, respectively, as compared with the values of 10.55 mA/cm^2, 0.68V and 4.0%, respectively, for a DSSC with an ordinary N3 dye [198].

Jiu *et al.* prepared nanocrystalline TiO_2 (anatase phase) with 3–5 nm in diameter with a surfactant–assisted sol–gel method. They assembled dye–sensitized solar cells using the prepared nano–scale TiO_2 crystals and reported their photocurrent–voltage performance. The results of this research reveal that the electrodes containing nano–scale TiO_2 crystals generate a significantly higher current density compared to the cell fabricated by big particles [199].

Highly–ordered TiO_2 nanotube arrays have been made by potentiostatic anodization of a titanium film in a fluoride containing electrolyte [200]. Grimes *et al.* [200, 201] described the application of this material architecture in both front–side and back–side illuminated dye–sensitized solar cells. The back–side illuminated solar cells are based on the use of 6.2 µm long (110 nm pore diameter, 20 nm wall thickness) highly–ordered nanotube–array films made through anodization of a 250 µm thick Ti foil in a KF electrolyte. Front–side illuminated solar cells use a negative electrode composed of optically transparent nanotube arrays, approximately 3600 nm in length, 46 nm pore diameter and 17 nm wall thickness. The arrays are grown on a fluorine doped tin oxide coated glass substrate by anodic oxidation of a previously deposited RF–sputtered titanium thin film in a HF electrolyte.

4.2. Hydrogen Production

The adverse effects of the air pollution on human health and global warming connected to green house gases call for the development of new environmental–friendly sources. Environmental and supply problems related to the use of fossil fuels have stimulated the search for alternative sources of energy. Hydrogen is a favored candidate, having a high calorific value and producing only water upon oxidation.

Solar rays in the near–infrared, visible and ultraviolet regions radiate a tremendous amount of energy toward the earth. Harnessing this solar energy would contribute significantly to our electrical and chemical needs. The application of photocatalysis to utilize such abundant and safe solar energy is much desired and vital to sustain life. Since the first energy crisis in the early 1970s, much research has been devoted to the development of efficient systems that would enable the absorption and conversion of solar light into useful chemical energy resources [202, 203]. One of the promising techniques to generate such "artificial photosynthesis" reactions is through the photocatalytic splitting of water using $NS–TiO_2$ to produce H_2 and O_2 under solar light [204, 205].

Since Fujishima and Honda [206] discovered the photocatalytic splitting of water into H_2 and O_2 on a TiO_2 photoanode, photocatalysis by TiO_2 has received much attention to investigate the photolysis of water. The mechanism of photocatalytic hydrogen production by TiO_2 is shown in Figure 12. When semiconductors are excited by photons with energy equal to or higher than their band gap energy level, electrons receive energy from the photons and are thus promoted from the valence band (VB) to the conduction band (CB). For TiO_2 semiconductor, the reaction is expressed by Eq. 3. The reduction and oxidation reactions are the basic mechanisms of photocatalytic hydrogen production and photocatalytic water/air purification, respectively. Both surface adsorption as well as photocatalytic reactions can be enhanced by nanosized semiconductors as more reactive surface area is available. For hydrogen production, the conduction band level should be more negative than the hydrogen production level ($E_{H2/H2O}$), while the valence band should be more positive than water oxidation level ($E_{O2/H2O}$) for efficient oxygen production from water by photocatalysis (see Figure 13). Theoretically, all types of semiconductors that satisfy the previously mentioned requirements can be used as photocatalysts for hydrogen production. However, most of the semiconductors, such as CdS and SiCn cause photocorrosion and are not suitable for water–splitting. Because of the strong catalytic activity, high chemical stability and long lifetime of electron/hole pairs, TiO_2 is the most widely used photocatalyst [22, 28, 203].

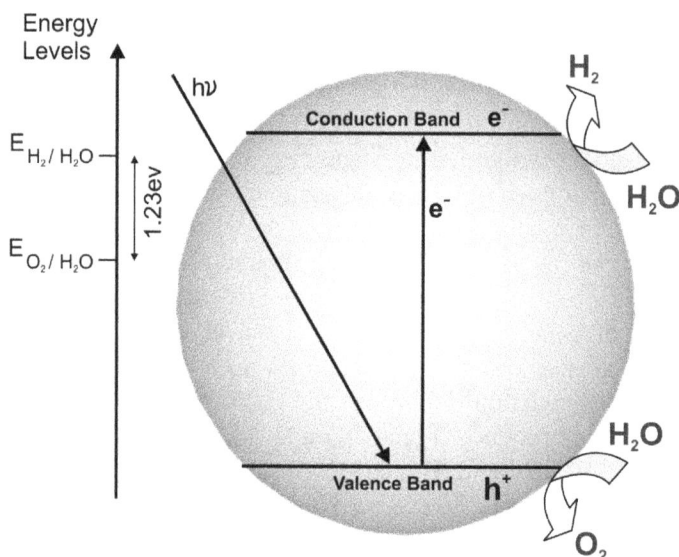

Figure 12. Mechanism of photocatalytic water–splitting for hydrogen production at the surface of TiO_2 nanoparticles.

Recently, interest in *NS–TiO$_2$* has been largely stimulated by the realization that the stability and chemical inertness of titanium dioxide are ahead of the materials so far considered as candidates for photoelectrodes in photoelectrochemical cells, which can use solar radiation as the energy resource [207–209].

Anpo *et al.* [202] provided a review paper on the advances achieved in the photocatalytic water splitting reaction, especially under visible light. In this report, they have focused on the preparation of visible light–responsive *TiO$_2$* thin films by a RF–magnetron sputtering method and its application for the separate evolution of H$_2$ and O$_2$ from water. They have also introduced various unique photocatalysts including cation or anion–doped metal oxides or metal oxynitride.

Ichikawa showed that hydrogen could be produced photocatalytically from natural seawater under sunlight at ambient temperature without the assistance of an external applied bias potential by using an optimized thin film of transparent titania photocatalyst [210]. Seawater has been considered as a naturally available electrolyte solution. Seawater has been used, without any further treatment, in the production of hydrogen in the presence of an optimized thin film of a photocatalyst (*TiO$_2$*). Sunlight was utilized as the source of energy to activate the photocatalyst (*TiO$_2$*) [210].

Sung Lee *et al.* [211] proposed the photocatalytic production of hydrogen from natural seawater with simple pre–treatments as demonstrated with La$_2$Ti$_2$O$_7$ suspension under ultraviolet light, CdS/TiO$_2$ suspension under visible light and Fe$_2$O$_3$ film under photoelectrochemical conditions. In all cases, the formation of harmful chlorine gas was not detected.

Nanosized *TiO$_2$* photocatalytic water–splitting technology has considerable potential for low–cost and environmentally friendly solar–hydrogen production [212, 213]. However, the rapid recombination of photo–generated electron/hole pairs as well as backward reaction and the poor activation of *TiO$_2$* by visible light are the main barriers [202, 203]. In response to these deficiencies, some investigators studied the effects of adding sacrificial reagents and carbonate salts to prohibit rapid recombination of electron/hole pairs and backward reactions [214, 215]. Other research focused on the enhancement of photocatalysis by modification of *TiO$_2$* by means of metal loading, metal ion doping, dye sensitization, composite semiconductor, anion doping and metal ion–implantation [216–218].

Bard *et al.* [219] reported the preparation of vertically grown carbon–doped TiO_2 ($TiO_{2-x}C_x$) nanotube arrays with high aspect ratios for maximizing the photocleavage of water under solar light irradiation. They found that the synthesized $TiO_{2-x}C_x$ nanotube arrays showed much higher photocurrent densities and more efficient water splitting under visible–light illumination (> 420 nm) than pure TiO_2 nanotube arrays. The total photocurrent was 20 times higher than that with a P25 nanoparticulate film under white–light illumination. According to their results, to maximize the water splitting efficiency of a TiO_2 photoanode, one would like:

(1) a narrower band gap to utilize visible–light energy,
(2) a high contact area with the electrolyte to increase the splitting of the electron/hole (e^-h^+) pairs and
(3) a thicker film to increase the total absorption of solar light.

Also, Misra *et al.* [220] designed a photoelectrochemical cell using carbon–doped titanium dioxide ($TiO_{2-x}C_x$) nanotube arrays as the photoanode and platinum. Pt nanoparticles incorporated in TiO_2 nanotube array, as the cathode (Figure 13). They found that photoelectrochemical cell was highly efficient (i.e. gave good photocurrent at a low external bias, j_p = 2.5–2.8 mA/cm^2 at –0.4 $V_{Ag/AgCl}$), inexpensive (only 0.4 wt % Pt on TiO_2) and robust (continuously run for 80 h without affecting the photocurrent) for hydrogen generation through water splitting under the illumination of simulated solar energy. The synthesis of the photoanode was carried out by the sonoelectrochemical anodization technique using aqueous ethylene glycol and ammonium fluoride solution. This anodization process gave self–organized hexagonally ordered TiO_2 nanotube arrays with a wide range of nanotube structures, which possessed good uniformity and conformability. In this work, as–synthesized titania nanotubes were annealed under a reducing H_2 atmosphere, which converted the amorphous nanotube arrays to photoactive anatase and it helped in doping the carbon (from the reduction of ethylene glycol) to give the $TiO_{2-x}C_x$ type photoanodes. The cathode material was prepared by synthesizing Pt nanoparticles (by reduction of a Pt salt to Pt(0)) into the titania nanotubular arrays through an incipient wetness method.

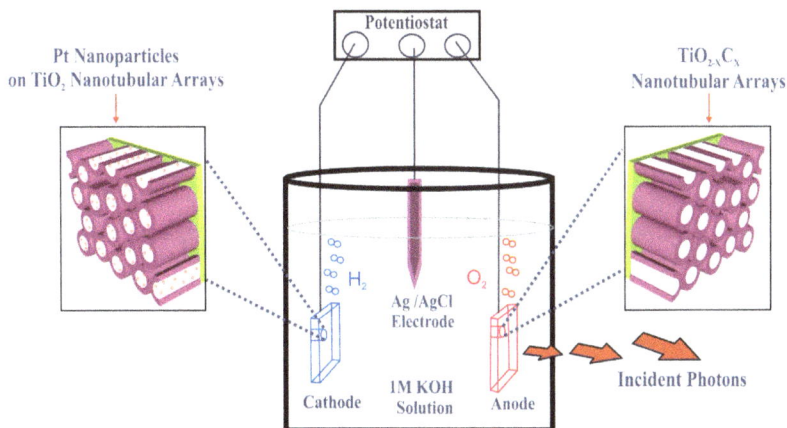

Figure 13. Schematic of the photoelectrolytic cell designed for the generation of hydrogen using a light source (UV or visible). The anode is carbon–doped titania nanotubular arrays prepared by the sonoelectrochemical anodization technique and the cathode is platinum nanoparticles synthesized on undoped titania nanotubular arrays. (Redrawn from Misra *et al.* [220] with permission from publisher, American Chemical Society. License Number: 2627061508363).

The previously mentioned reaction system for the production of H₂, using photocatalytic splitting of water under visible light irradiation, has an obvious practical disadvantage. The photocatalytic splitting of water always into produces a gas mixture. This calls for a separation process of the gases before H₂ can be effectively utilized. Construction of a photocatalytic system enabling the separation of H₂ and O₂ from water under visible light irradiation is, therefore, of vital interest. In order to achieve this goal, Anpo *et al.* have reported that the separation of H₂ and O₂ could be achieved by visible light–responsive *TiO₂* thin films from water under visible light irradiation by applying an H–type glass container, as shown in Figure 14 [221–223]. The *TiO₂* thin film device (vis–*TiO₂*–Ti/Pt) consists of the Ti foil with 50 μm thickness deposited on one side with vis–*TiO₂* thin film and on the other side with Pt. The prepared *TiO₂* thin film device is mounted on an H–type glass container (Figure 14), separating the two aqueous solutions. The *TiO₂* side of the thin film device is immersed into 1.0 N NaOH aqueous solution and the Pt side is immersed into 1.0 N H₂SO₄ aqueous solution in order to add a small chemical bias between the two aqueous solutions. The visible light irradiation ($\lambda > 450$ nm) of the vis–TiO_2–Ti/Pt mounted in the H–type glass container leads to the stoichiometric evolution of H_2 and O_2 separately with good linearity against the irradiation time [223].

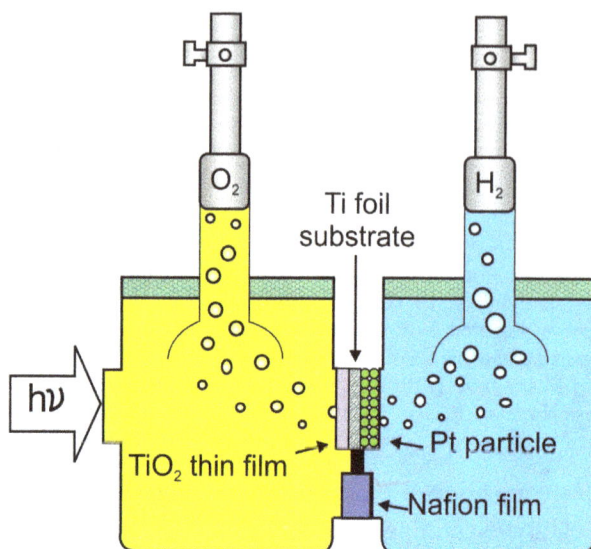

Figure 14. H–type glass container for the separate evolution of H_2 and O_2 using a TiO_2 thin film device.

4.3. Hydrogen Storage

Hydrogen is widely considered as a strategic energy carrier, in particular if new energy sources will be developed in the future to reach a minor dependence on petroleum. Compared to fossil fuels, hydrogen offers significant advantages, in particular the highest energy density (35.7 kW h/kg) and the absence of carbon atoms, which implies zero CO_2 emissions in the oxidation reaction products [224, 225]. However, the successful implementation of the hydrogen economy into industry raises the challenges of storage and transportation of hydrogen. Recent approaches to hydrogen storage have considered adsorption of hydrogen on solids of large surface area [226], hydrogen storage by metal hydrides [227] and intercalation of molecular hydrogen in clathrate hydrates [228].

The recent discovery of hydrogen clathrate hydrate, $(32+x)H_2 \cdot 136H_2O$ opens up the possibility to incorporate molecular hydrogen into the cage of water molecules such that dissociation of hydrogen can be avoided and high uptake of hydrogen can be achieved [229]. In such a clathrate hydrate, the molecule of hydrogen is stabilized by several OH groups through hydrogen bonding [230, 231]. It is necessary to apply an extremely high pressure to hydrogen to promote self–organization of water and introduce hydrogen molecules into the clathrate structure. A breakthrough can be anticipated by the use of a preformed "host" structure to accommodate the hydrogen "guest" molecules. Such host structures should have several OH groups and cavities of suitable geometry where the pore diameter is larger than the dynamic diameter of a free hydrogen molecule (d_1 = 0.4059 nm). A possible candidate for such "host" structures is multilayered TiO_2 nanotubes which have multilayered walls [232]. The interstitial spacing between layers (d_2 = 0.72 nm) contain ion–exchangeable OH groups which could accommodate hydrogen molecules. The size of interstitial cavities–zigzag slit pores formed between two (100) planes is ca. 0.72 nm. This is larger than the dynamic diameter of hydrogen molecules (0.41 nm) and greater than the nuclear distance in the hydrogen molecule (0.07 nm). OH groups in the nanotube lattice could stabilize the hydrogen molecules via weak van der Waals interactions resulting in the formation of $TiO_2.xH_2$ clathrates [28, 225, 233, 234].

Formation of vertically oriented and ordered–TiO_2 nanotubes through different processes has been reported by several research groups [235–245]. The ability of TiO_2 nanotubes to reversibly accumulate molecular hydrogen in great concentrations over a wide range of temperatures ranging from −195 to 200°C, [225, 235] opens up the possibility for hydrogen storage and related applications [246–251].

Walsh *et al.* studied the reversible storage of molecular hydrogen by sorption into multilayered TiO_2 nanotubes prepared by hydrothermal process. In this research, the sorption of hydrogen between the layers of TiO_2 was studied under temperatures between −195 to 200°C and at pressures of 0 to 6 bar. Hydrogen could intercalate between layers in the walls of TiO_2 nanotubes forming host–guest compounds $TiO_2.xH_2$, where x≤1.5 and decreases at higher temperatures. The rate of hydrogen uptake increased with temperature and the characteristic time for hydrogen sorption in TiO_2 nanotubes was several hours at 100°C [225].

The formation of vertically oriented nanotubular TiO_2 arrays has also been reported by a simple anodization process [246]. The major advantage of the anodization method to prepare $NS–TiO_2$ is its ability to scale–up. As the nanotubes are vertically oriented and form a good electrical contact with the metal substrate, well–controlled functionalization of the nanotubes can be easily achieved. By appropriate cathodic pulsing cycles, the nanotube array can be extracted from the Ti substrate or a thin sputtered layer of Ti can be completely anodized to form a thick TiO_2 nanotubular array. Hydrogen storage studies have been carried out on TiO_2 nanotubular arrays having different diameters (30, 50 and 100 nm) by charging and discharging hydrogen with potentiostatic/galvanostatic control [246].

Lin *et al.* found that TiO_2 nanotubes can store up to ~2 wt % H_2 at room temperature and 6 MPa pressure. However, only about 75% of this stored hydrogen could be released when the hydrogen pressure was lowered to ambient conditions due to physisorption. Approximately 13% was weakly chemisorbed and could be released at 70°C as H_2 and approximately 12% was bonded to oxide ions and released only at temperatures above 120°C as H_2O molecules. Their results indicated that at room temperature and a pressure of ~900 psi (6 MPa), the atomic ratio of H/TiO_2 was ~1.6, corresponding to ~2.0 wt % H_2 for TiO_2 nanotubes, compared to a much lower hydrogen concentration of ~0.8 wt % for bulk TiO_2. The group also found that when the pressure was reduced, only ~75% of the stored hydrogen could be released. The other 25% hydrogen molecules were retained due to chemical adsorption [247].

Antonelli *et al.* studied the hydrogen storage in the mesoporous and microporous of Ti oxides, synthesized from C_6, C_8, C_{10}, C_{12} and C_{14} amine templates possessing BET surface areas ranging from 643 to 1063 m^2/g. They found that at 77 K the isotherms for all materials gently rose sharply at low pressure and continued to rise in a linear fashion from 10 atm to 65 atm and then return on desorption without significant hysteresis. Extrapolation to 100 atm could yield total storage values as high as 5.36 wt % and 29.37 kg/m^3 and surface Ti reduction by the appropriate organometallic reagent providing an increase in performance [248].

From the above studies, it can be concluded that in contrast to carbon nanotubes or metal–alloy hydrides, $NS–TiO_2$ materials, especially titanate nanotubes, can also operate over convenient ranges of pressure and temperature. Moreover, simple pressure and temperature swings can be used to adsorb and desorb hydrogen from the solid–state, nanotubular titanates. Such a selectiveness of titanate nanotubes for sorption of hydrogen can also be used in the design of the membranes for the separation of hydrogen from other gases. This could find an application in various industrial processes, such as the water–shift reaction.

4.4. Sensors

There is a general opinion in both scientific and engineering communities urges for the development of cheap, reliable sensors to control and measure systems as well as automate services and microelectronics with reasonable reliability performance and low price [252]. For the development of sensors, interest has increased to study the transduction principle, simulation of systems and structural investigations of the materials and choice technology. In many aspects of today's life, the use of gas sensors becomes increasingly important. These devices are not suited to make high precision measurements of gas concentrations. They were designed to detect the presence of target gases and give a warning if given threshold values are attained. Binary n–type semiconductor oxides such as tin oxide (SnO_2) [253], indium oxide (In_2O_3) [254] or zinc oxide (ZnO) [255] have been extensively investigated as gas–sensing materials. These metal oxide sensors detect small amounts of a gaseous species present in the air from a change in electrical resistance. In the first years of developing oxygen sensors for automotive exhaust gases, semiconductor oxides like TiO_2 were considered as a good alternative to zirconia. Many studies were carried out with titania, especially by Ford Motors [256, 257]. In addition, TiO_2 nanocrystalline films have been widely studied as sensors in gas and liquid phases by different research groups [258–270].

Francioso *et al.* [257] tested a sol–gel synthesized *NS–TiO₂* thin film sensor for automotive applications. The experimental trials were carried out to control exhaust gas after combustion in spark ignition engines. The sensor responses have been successfully acquired in a controlled environment and on a gasoline engine bench. Results were satisfactory and times of response were fast and different as compared to other lean and rich mixtures. The comparison with commercial lambda probe showed a good time response and a good correlation with *NS–TiO₂* sensor. Regarding single component of mixture of flue gas, higher responses were obtained for oxygen and nitrogen oxide, rather than CO_2 and CH_4. Furthermore, the $NS-TiO_2$ sensor has a better thermal conductivity than Bosch Lambda probe having no problem of slow activation [257].

Gas selectivity is a very important characteristic that measures the ability of a sensor to precisely identify a specific gas. This feature is necessary to develop integrated gas sensor arrays. Hydrogen is an important chemical in many industrial processes. However, it leaks easily from systems. It is dangerous because hydrogen is an explosive gas. Therefore, a lot of effort has been put into investigating hydrogen sensors and to improve their selectivity [268, 271–274]. Wang *et al.* [272] prepared a new highly selective H_2 sensor based on TiO_2/PtO–Pt dual–layer films. The results indicated that at 180–200°C, the prepared nanostructured sensor exhibited good sensitivity to H_2 in air immune to many other kinds of reductive gases (e.g. CO, NH_3 and CH_4). The group found that the sensor could give a faithful response to 1% H_2 in the air, while the limitation for detecting H_2 in nitrogen was less than 1000 ppm.

Taurino *et al.* conducted study on gas sensing properties of $NS-TiO_2$ thin films grown by seeded supersonic beam of cluster oxides. The group indicated successful application of supersonic cluster beams produced by a pulsed micro–plasma cluster source in the preparation of nanocrystalline thin films of TiO_2. Sensors showed a good response to ethanol, methanol and propanol [275]. Recently, Taurino and co–workers achieved very interesting gas sensing results by using a thin layer of TiO_2 nanoparticles deposited by a matrix assisted pulsed laser evaporation (MAPLE). Electrical tests performed in a controlled atmosphere in the presence of ethanol and acetone vapors releave a high value of the sensor response even at very low concentrations (20–200 ppm in dry the air) for both vapors. A higher response and a higher sensitivity were achieved for ethanol as compared with acetone. Based on resistive transduction mechanism in the detection of ethanol and acetone at low air level, MAPLE TiO_2 gas sensors are considered promising [276].

$NS-TiO_2$ materials have been used for ammonia detection [277–280]. Suh *et al.* [277] proposed a thin film gas sensor planar structure of fabricated with TiO_2 to monitor ammonia. They deposited a thin sensitive TiO_2 film by a DC reactive magnetron sputtering technique onto a cleaned silicon substrate equipped with interdigitated comb shaped electrodes. A static gas sensing mechanism has been employed to analyse the sensing ability of the prepared sensors. Their results revealed that as–deposited films were not sensitive to the ammonia gas. However, films annealed at 873 K, with good crystallinity were found to exhibit a good sensing property. The selectivity for ammonia gas was highest sensitivity at an operating temperature of 250°C. Response and recovery times of this sensor for a flow of 500 ppm of ammonia were evaluated as 90 and 110 s, respectively.

Xiao *et al.* [281] reported H$_2$O$_2$ sensor based on the room–temperature phosphorescence of nano *TiO$_2$/SiO$_2$* composite. A *TiO$_2$/SiO$_2$* composite was prepared by sol–gel method. Their results showed that this nanocomposite could produce a highly emissive broadband at room–temperature phosphorescence from 450 to 650 nm at an excitation wavelength of 403 nm. The white phosphorescence of *TiO$_2$/SiO$_2$* could be quenched by H$_2$O$_2$. The phosphorescence quenching effect demonstrated reasonable sensitivity and high selectivity to H$_2$O$_2$. It was also successful in establishing a new sensor of high sensitivity, high reproducibility, fast response and high selectivity to H$_2$O$_2$.

One–dimensional (1D) *TiO$_2$(B)* nanowires have been synthesized via a facile solvothermal route and applied as humidity sensors [282, 283]. The synthetic *TiO$_2$(B)* nanowire electrode exhibited unique electronic properties, e.g., favorable charge–transfer ability, negative–shifted appearing flat–band potential, existence of abundant surface states or oxygen vacancies and high–level dopant density. Moreover, the obtained *TiO$_2$(B)* nanowires were found to display good humidity sensing abilities as functional materials in the humidity sensor application. With relative humidity increasing from 5% to 95%, about one and a half orders of magnitude change in resistance was observed in the *TiO$_2$(B)* nanowire–based surface–type humidity sensors [282].

Chemical oxygen demand (COD), which represents the total pollution load of most wastewater discharges, is a main index to assess the organic pollution in aqueous systems. Typically, for a COD determination, the organic compounds presented in the water sample are oxidized completely by an added strong oxidant, usually K$_2$Cr$_2$O$_7$ or KMnO$_4$. The index is calculated by determining the quantity of the consumed oxidant and expressing it in terms of its oxygen equivalent. To achieve the complete oxidation of the organic pollutants, it is necessary to introduce excess oxidants and heavy metal salts serving as the catalyst

[284, 285]. This operation increases both the cost and the risk of water pollution. Simultaneously, the conventional COD determination method requires a long analysis time, which hinders its application in environmental assessments. $NS–TiO_2$ photocatalytic sensors have been utilized to determine COD in water research [286, 287]. Experimental data indicates that the photocurrent of the $NS–TiO_2$–based sensor changes linearly with COD amount in the range of 0.5–235 mg/L. The method possesses many advantages such as simplicity of preparation, low cost of manufacturing process for the sensor, fast response time, acceptable lifetime and potential for automated monitoring. However, there are still several factors limiting the method from a wide application in water assessment. For example, a narrow linear range and liability to be influenced by the reductive or oxidative substances presented in the sample, such as O_2, chloride and S^{2-} [287].

The use of TiO_2 thin film for oxygen sensing has been investigated extensively [262, 298–303]. Oxygen sensors based on $NS–TiO_2$ materials include $CeO_2–ZrO_2–TiO_2$ [289], TiO_{2-x} [290], $SnO_2–TiO_2$ [291, 292], $Cr_2O_3–TiO_2$ [293], $V_2O_5–TiO_2$ [294], Cr– [292, 295, 296], Nb– [292, 296, 297] and Pt–doped $NS–TiO_2$ [262, 288, 298]. Elyassi *et al.* reported the development of an oxygen sensor for automotive applications with a solid–state reference by employing a ternary mixed oxide of $CeO_2–ZrO_2–TiO_2$. The group found that unlike conventional sensors, where the developed voltage varies between 1000 and 100 mV, the new $NS–TiO_2$–based sensor exhibited a narrower variation in the voltage ranging from +300 to –250 mV [289].

Lee and Hwang [291] fabricated SnO_2/TiO_2 thin films on SiO_2/Si and Corning glass 1737 substrates using RF magnetron sputtering process. They carefully examined the gas sensing properties of these films under an oxygen atmosphere with and without UV irradiation. It was determined that the oxygen sensitivity of the films deposited on Corning glass 1737 substrates was significantly lower than that of the films grown on SiO_2/Si substrates. According to group's findings, when a SnO_2/TiO_2 coupled–thin–film is exposed to oxygen gas, oxygen molecules are captured by the surface electrons and become adsorbed oxygen (O^-_{ads}), i.e.

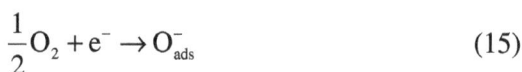

$$\frac{1}{2}O_2 + e^- \rightarrow O^-_{ads} \tag{15}$$

The adsorbed oxygen creates a depletion layer on the surface of the film and increases the energy barrier, thereby increasing the electrical resistance of the film. The significant changes observed in the resistance of the sensors can be explained by reference to the model shown in Figure 15, in which electrons modulate the depletion region. For gas sensor applications, the larger the variation in the resistance, the higher the achieved sensitivity [291, 299].

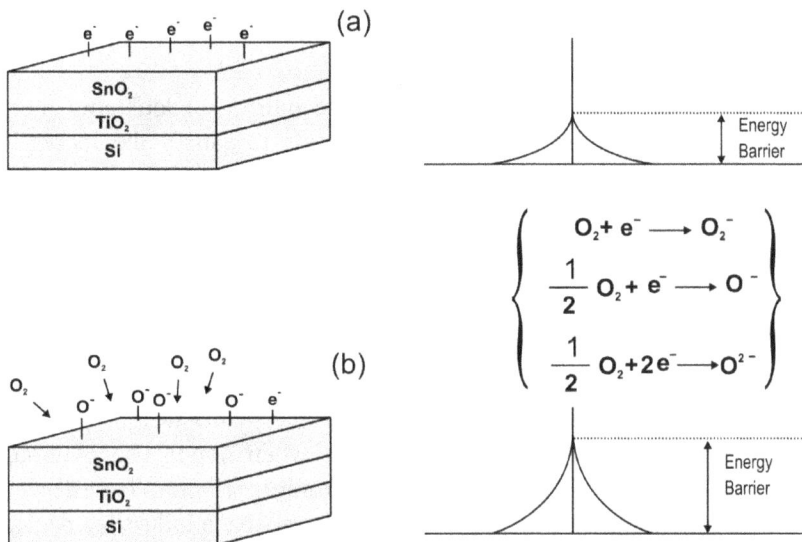

Figure 15. Schematic illustration of sensing mechanism of SnO$_2$/TiO$_2$ thin films: (a) SnO$_2$/TiO$_2$ film in vacuum and (b) tin oxide film exposed to oxygen. In Fig. 15(b), oxygen molecules are adsorbed and receive electrons creating a depletion layer (O$^-$) on the surface of the SnO$_2$/TiO$_2$ thin film. The depletion layer increases the resistance of the film. (Redrawn from Lee and Hwang [291] with permission from publisher, Elsevier. License Number: 2627100491014).

Zakrzewska found that titanium dioxide doped with Nb and Cr should be considered as a bulk sensor. Its performance was governed by the diffusion of point defects, i.e. very slow diffusion of Ti vacancies for *TiO$_2$*: 9.5 at% of Nb and fast diffusion of oxygen vacancies in the case of *TiO$_2$*: 2.5 at% Cr sensor. The corresponding response times were 55 min for *TiO$_2$*: 9.5 at% of Nb and 20 s for *TiO$_2$*: 2.5 at% Cr [292].

Pt–doped *TiO$_2$* sensors showed low operation temperature (350–800°C), improved gas sensitivity and short response time (<0.1 s) [300, 301]. The oxygen–sensing mechanism is a combination of Pt/*TiO$_2$* interfaces in a Schottky–barrier mechanism and an oxygen vacancy bulk effect mechanism [302]. It has been demonstrated that at high temperatures,

TiO_2 devices could be used as thermodynamically controlled bulk defect sensors to detect oxygen over a large range of partial pressures. At low temperatures, Pt/TiO_2 Schottky diodes made extremely sensitive oxygen detection possible [298, 303].

4.5. Batteries

One of the important challenges in the 21st century is undoubtedly energy storage. It is essential to seek new materials to satisfy the increasing demands for energy conversion and storage worldwide. Nanostructured materials such as nanosized TiO_2 are attractive for energy devices [304, 305]. In this section, we describe the potential application of $NS-TiO_2$ materials in rechargeable batteries like lithium–based batteries.

Lithium–ion batteries, referred to as Li–ion batteries, are a type of rechargeable batteries commonly used in consumer electronics. They are currently one of the popular types of batteries for portable electronics. They are characterized by one of the best energy–to–weight ratios, no memory effect and a slow loss of charge when not in use. Although originally intended for consumer electronics, Li–ion batteries are growing in popularity with the defense and aerospace industries because of their high energy density [306–309].

$NS-TiO_2$ has attracted much attention in Li–ion batteries insertion because it is both a low voltage insertion host for Li and a fast Li insertion/extraction host. These characteristics render $NS-TiO_2$ a potential anode material for high–power Li–ion batteries. Nanostructured titanates also attract attention as a possible negative electrode for rechargeable lithium batteries owing to their open, mesoporous structure, efficient transport of lithium ions and favored ion–exchange ratio that results in a high value of charge/discharge capacity, good kinetic characteristics and good robustness and safety [310–313]. These new $NS-TiO_2$–based electrodes can substitute commercial carbon negative electrode batteries, which suffer from safety concerns [28, 314]. Titanates are also of interest as anodes in hybrid supercapacitors, where they are combined with a carbon positive electrode and a non–aqueous electrolyte [315, 316]. The electrode reaction on nanostructured electrodes includes intercalation of lithium ions according to the following reaction:

$$TiO_2 + xLi^+ + xe^- \leftrightarrow Li_xTiO_2 \qquad (16)$$

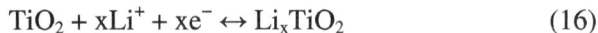

where x is the lithium insertion coefficient [28, 317].

Recent studies on the use of *NS–TiO$_2$* (anatase) as anode with LiCoO$_2$ cathode in a lithium battery demonstrated specific capacity of 169 mAh g^{-1}. The Li insertion was limited to a maximum x value of 0.5 in equation 16 [317]. This can be compared with Li$_{0.91}$TiO$_2$(B) corresponding to a capacity of 305 mAh g^{-1} at a potential of 1.6V versus Li$^+$ (1 M)/Li for *TiO$_2$(B)* nanowires (20–40 nm diameter) [318].

Xu *et al.* studied electrochemical properties of hydrothermally synthesized anatase *TiO$_2$* nanotubes with diameters of about 10 nm and lengths of 200–400 nm, as an anode material for Li–ion batteries. There were long voltage plateaus in the discharging/charging curves and the lithium insertion capacity at the potential plateau of 1.73 V in the first cycle was about 150 mAh g^{-1}. They found that the discharge capacity still kept at 210 mAh g^{-1} in the 30th cycle, presenting a good high–rate cycling performance [312].

Yang *et al.* investigated electrochemical performance and spectroscopic characterizations of the decomposition products from electrolytes on native and *TiO$_2$*–coated LiCoO$_2$ and LiMn$_2$O$_4$ in different potential regions. They found that *TiO$_2$*–coated materials exhibited better cyclic stability in different potential regions (i.e., 3.0–4.3 V and 3.0–4.6 V) and the decomposition of the electrolytes was suppressed on the coated materials surface. This was attributed to the presence of *TiO$_2$* on the surface to separate active particles from the organic electrolyte [310].

Ag–modified *TiO$_2$* nanotube was also used as anode material for Li–ion battery. *TiO$_2$* nanotubes prepared by using a hydrothermal process were first coated with silver nanoparticles as the anode material for Li–ion batteries by the traditional silver mirror reaction. It was found that, because of the high electronic conductivity of metal Ag, Ag additive significantly improved the reversible capacity and the cycling stability of the *TiO$_2$* nanotube at a high charge–discharge rate and decreased the cell polarization [313].

Mechanically blended composite of nanosized *TiO$_2$* and carbon nanotubes (CNTs) has also been used as potential anode materials for Li–ion batteries. It was found that the *TiO$_2$*/CNTs nanocomposite exhibited an improved cycling stability and higher reversible capacity than CNTs. The reversible capacity of the *TiO$_2$*/CNTs composite reached

168 mAh g^{-1} at the first cycle and remained almost constant during long–term cycling [315, 316].

The current tendency in research is to switch from nanotubular to nanofibrous titanate nanostructures in lithium battery applications. The latter structure retains better capacity retention on cycling [319, 320]. Nanofibrous titanates include protonated titanate nanofibers [320] produced by alkaline hydrothermal treatment and acid washing, spinel $Li_4Ti_5O_{12}$ nanofibers [321] produced from protonated titanate nanofibers by hydrothermal ion exchange, nanofibers of $TiO_2(B)$, [37, 40, 121] and titanate nanofibers calcined at 400°C [322]. The lithium insertion coefficient, x varies from 0.71[320] to 0.91[318] for nanofibrous structures. The typical cyclic voltammogram of titanate nanofibers indicated several pairs of voltage peaks in the range of 1.5 to 2.0 V relative to Li$^+$/Li [28, 319, 323].

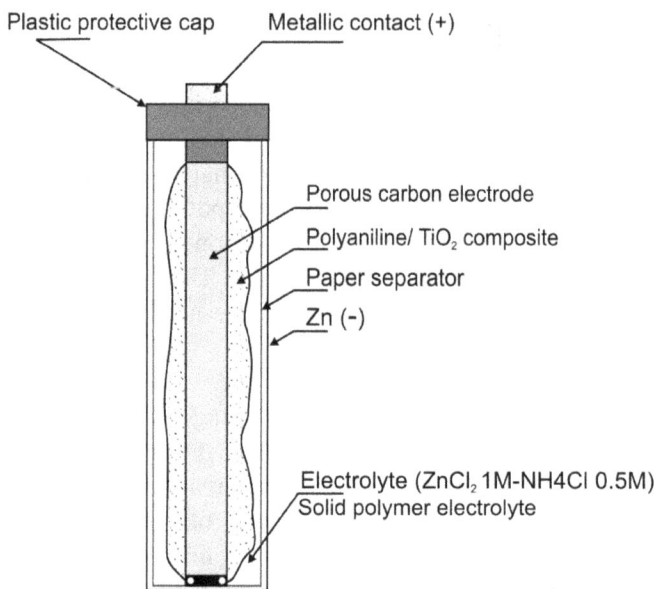

Figure 16. Battery cell fabrication by using polyaniline/TiO_2 composite as the cathode material. (Adapted from Gurunathan *et al.* [324] with permission from publisher, Elsevier. License Number: 2627061090272).

Polyaniline/TiO_2 composite has been used as a cathode material for rechargeable batteries comprising zinc container as the anode, cellulose acetate as the separator and polyvinyl sulfate and carboxy methyl cellulose as the solid polymer electrolytes (SPE) (Figure 16) [324]. The open circuit voltage of this rechargeable battery is 1.4 V and current is 250 mA to 1.0 A, for 50 recharge cycles. The power density is 350 A h/kg and power efficiency is 70%. The advantages of this kind of rechargeable battery are at least 50 rechargeable cycles, 25% less weight and low production cost [324].

4.6. Cancer Prevention and Treatment

We are at the threshold of a new era in cancer treatment and diagnosis, brought about by the convergence of two disciplines. The innovative field of nanobiotechnology is developing many drug delivery concepts that promise real–life applications. Nanoparticles, engineered to exquisite precision using polymers, metals, lipids and carbon have been combined with molecular targeting, molecular imaging and therapeutic techniques to create a powerful set of tools in the fight against cancer [325]. Application of $NS–TiO_2$ in photocatalytic process indicated a potential in treating cancer through photodynamic therapy (PDT). Photodynamic therapy involves injecting the tumor with PDT drugs that upon radiation react with oxygen in the vascular system to produce singlet oxygen [326].

The effect of TiO_2 nanoparticles on experimental animals has been investigated to determine the possibility of induced toxicity. These experiments consistently yield negative results. For example, when TiO_2–coated mica particles were administered to rats in their diet for up to 130 weeks no significant changes in body weight gains, or hematological or clinical chemistry parameters were observed [327].

Fujishima *et al.* studied the tumor cell killing effect of TiO_2 nanoparticles (average diameter of 30 nm) with UV irradiation. A distinct HeLa cell killing effect was observed *in–vitro* with photoexcited TiO_2 nanoparticles and tumor growth of HeLa cells was significantly inhibited by the treatment of TiO_2 nanoparticles with UV irradiation [4, 327]. According to this report, HeLa cells initially subcultured *in–vitro* were injected into the backs of nude mice. When the tumor became measurable in size (about 2 weeks after inoculation of the cells), the tumor–bearing mice were divided into four groups of five mice each. TiO_2 nanoparticles were injected into the tumor and surrounding subcutaneous tissue. Three days after the TiO_2 injection, the skin covering the tumor was opened surgically and the tumor was irradiated directly by mercury lamp for 1 h (300–400 nm). The skin was then closed. The tumor size was measured at 2– or 3–day intervals. According to the experimental results and observations the treatment clearly inhibited the tumor growth.

Water–soluble single–crystalline TiO_2 nanoparticles have also been developed for photocatalytic cancer–cell treatment [328]. Cheon *et al.* [328, 329] have presented the fabrication of water–soluble and biocompatible TiO_2 nanoparticles. These were anatase–phase single–crystalline particles and their photocatalytic capabilities for treatment of skin cells were examined. In this work, TiO_2 nanoparticles have been synthesized by a high–temperature nonhydrolytic method [329]. As synthesized, the nanoparticles were insoluble in water due to hydrophobic capping ligands. In order to make them water soluble, TiO_2 nanoparticles were coated with 2,3–dimercaptosuccinic acid (DMSA). The coated TiO_2 nanoparticles became easily redispersed in water without any noticeable aggregation. Furthermore, these showed fairly good colloidal stability within a tested NaCl concentration of 250 mm and in the pH range of 6 to 10 [328, 329].

Woloschak *et al.* [330] developed the intracellular use of a new type of bio–nanocomposite that had novel functional properties inside cells and *in–vitro*. These nanocomposites were consisted of TiO_2 nanoparticles (4.5 nm in size, surface coated with glycidyl isopropyl ether) and DNA oligonucleotides bond via dopamine to the nanoparticles. Within the nanocomposites DNA oligonucleotides retained base–pairing specificity, TiO_2 nanoparticles exhibited photoreactivity. TiO_2 nanocomposites demonstrated semiconducting properties through both constituents– excitation of TiO_2 (exposure to electromagnetic radiation of energy above 3.2 eV) resulting in charge separation with accumulation of electrons in metal oxide and irreversible trapping of the electropositive holes in the sugar molecules of the DNA phosphodiester backbone leading to the cleavage of the DNA. Woloschak and co–workers cleaved the DNA to remove the TGF–beta receptor from $CD8^+$ T cells for the treatment of prostate cancer in a mouse [331, 332].

The use of nanostructured reservoirs for drug delivery has been focused mainly to the treatment of neurodegenerative diseases such as epilepsy or cancer tumors [333–335]. In this way, cancer is a disorder of the processes of cellular growth, development and repair. It consists in an uncontrolled growth of cells that differ morphologically and biochemically from the original ones. Tumors of the central nervous system (CNS) are the most frequently found in children (70–80%) and they arise from glial cells tending to metastasize outside the CNS unless there is a surgical intervention. Chemotherapy plays an important role in the treatment of both recurrent and newly diagnosed patients. One of the important means to control and treat these CNS tumors is through the use of Temozolomide. This is an efficient drug active in the control of high grade brain gliomas whose complete surgical removal is seldom accomplished because of their infiltrative nature [336, 337]. A matrix of $NS–TiO_2$ has been synthesized by sol–gel process to immerse temozolomide. From the results of this research, it can be determined that the $NS–TiO_2$ matrix reservoir has the ability to be obtained with the temozolomide occluded in it, however the conditions of synthesis must be considered carefully.

In general, the best way to eliminate a problem is to eliminate the cause. In cancer, the problem can be perceived differently at various stages of the disease. Most apparently, if genetic mutations are the underlying cause, then we must counteract the causes of the mutations. Unfortunately, genetic mutations are caused by artificial or natural carcinogens only part of the time. At other instances, they may occur spontaneously during DNA replication and cell division. With present science and technology there is very little we can do to prevent this from happening. However, in all other cases, eliminating the carcinogens is indeed a highly effective way in cancer prevention. But most patients do not recognize the problem until it has actually occurred, which makes preventive medicine a rarely thing, although a highly effective form of cancer prevention [338].

The question is, is there a way to eliminate cancer through nanostructured materials like $NS–TiO_2$ before it starts? To demonstrate the viability of the $NS–TiO_2$–based treatments, let us consider melanoma for example. Melanoma, a form of skin cancer, is caused primarily by ultraviolet radiation from the Sun (Figure 17). Solar rays are known to cause irreversible photodamage to the skin. The biologically important output from the sun reaching the earth's surface contains UV, visible and infrared wavelengths. Although UV radiation (180–400 nm) is the smallest fraction (~15%) of terrestrial solar radiation, it is the most energetic and induces damage to living systems. UV radiation is divided into three types, namely UVC (180–280 nm), UVB (280–320 nm) and UVA (320–400 nm). Virtually all UVC and 99% of UVB are filtered out by the earth's ozone layer but about 99% of UVA reaches the earth's surface. Hence, UVA makes up 90–95% of the total UV radiation that reaches the surface of the earth. Indeed, the amount of solar UVB and UVA reaching the earth's surface is affected by latitude, altitude, season, time of the day, cloudiness and the quality of the ozone layer. UV radiation–induce skin effects include acute responses such as sunburn, pigmentation, hyperplasia, immunosuppression and vitamin D synthesis and chronic effects, like photocarcinogenesis and photoaging. UVA wavelengths penetrate more deeply into the skin and are, therefore, more responsible than UVB for photoaging, wrinkling of the skin and the formation of skin cancer. Studies indicate that about 90% of sunlight's melanoma causing effect may come from UVA rays [338–340]. Reactive oxygen species ($\cdot O_2^-$, $\cdot OH$ and $HO_2^{\cdot-}$) are produced from molecular oxygen during mitochondrial respiration in the normal metabolism of aerobic cells. To protect them from oxidative injury, aerobic cells have evolved a complex antioxidant defense system. However, if the reactive oxygen species load overwhelms the antioxidant defense, damage to cellular components, including DNA, proteins, lipids and membranes eventually occurs. It is well known that UV radiation induces reactive oxygen species formation in cutaneous tissues and UV radiation–induced reactive oxygen species damage DNA, mitochondria and induce apoptosis in cell cultures [339, 341].

The current method of preventive treatment against bombardment with this kind of harmful radiation involves suspending a substance that either absorbs or scatters ultraviolet radiation in a thick emulsion. We use this emulsion, called sunscreen, to coat our skin prior to prolonged exposure to sunlight [338]. Sunscreens typically contain 'chemical filters', that are organic compounds that absorb strongly the UV (most often, UVB) and 'physical filters', such as TiO_2 and ZnO that block UVB and UVA sunlight through absorption, reflection and scattering [339].

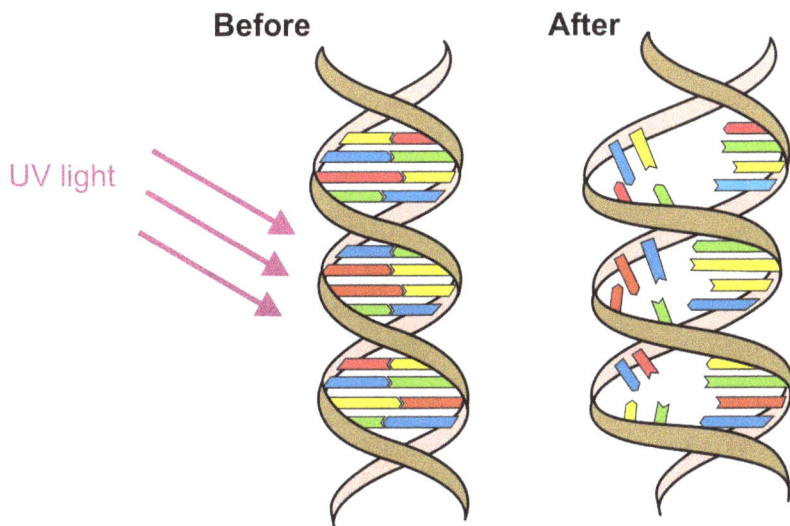

Figure 17. UV radiation is one of the most prominent causes of DNA damage. Since UV radiation is high frequency and thus high energy, it can easily damage the delicate DNA double helix. Individual nucleotide bases readily absorb UV radiation and can become excited after even short–term exposure. This can break hydrogen bonds between the two complementary chains and sometimes even the covalent bonds between the phosphate backbone and the ribose, causing genetic mutations. The result is a mutated genetic sequence and the production of defective proteins.

The inorganic ingredients of sunscreens, titanium dioxide and zinc oxide, are among the most efficient UVA blockers in sunscreens [342]. These compounds have been used for years in consumer products and are generally considered to be inert. However, TiO_2 is susceptible to excitation by UV radiation because it is a semiconductor with a band gap in the region of 3.2 eV, corresponding to UV wavelengths below 390 nm [46]. Photoexcited TiO_2 can generate a number of reactive oxygen species in aqueous solutions and there is evidence that use of sunscreens containing microfine TiO_2 (30–220 nm) can result in the percutaneous absorption of titanium [339]. Because encapsulation of TiO_2 within zeolites alters its photocatalytic activity, supramolecular composites based on NaY zeolite hosts containing TiO_2 guests were prepared and the effects on reactive oxygen species formation in cells under UVA–irradiation evaluated by Ann English *et al.* [339]. Protection of cells against TiO_2–induced intracellular reactive oxygen species by encapsulation suggests that supramolecular species may be beneficial in photoprotection of the skin.

Another problem with using sunscreens is that this emulsion can be easily rubbed off and can loose its effectiveness over time, thus needing to be reapplied periodically. An even bigger problem is that we leave openings in the sunscreen coating during sunscreen application due to macro–scale and micro–scale imperfections in our skin. This allows UV radiation to permeate through the dead layer of skin, spreading out to a wider area due to slit diffraction and causing more widespread damage. All of these problems take away from the overall effectiveness of this preventive method.

Recent studies have indicated that it is possible to tag specific types of cells with nanoparticles by conjugating them to targeting agents designed to recognize cell–specific surface proteins [338]. Nanoparticles attach to drugs or substances can be conjugated to short peptide chains, proteins or artificial nanobodies. If we manufacture nanoparticles attached to UV scattering substances like titanium oxide and zinc oxide, or UV absorbing substances like octyl methoxycinnamate and oxybenzone, to target skin cell surface proteins, it may possible to coat these cells with sunscreen on the nanoscale. This nanotechnology–based preventive treatment method may eliminate most of the problems mentioned above. If the cells can be coated directly, the problem of diffraction in areas that are sparsely coated can be eliminated.

4.7. Antibacterial and Self–Cleaning Applications

One of the most important applications of $NS–TiO_2$, which has been commercialized recently, is antibacterial and self–cleaning tasks. Several of these applications are listed in Table 5.

In the section below we incorporate a detail discussion regarding $NS–TiO_2$ as agent in self– cleaning and antibacterial applications.

$NS–TiO_2$–based self–cleaning products (e.g. cement, concrete, tiles, glass and tents) have been widely utilized in all over the word. Such products can be maintained clean by the combined action of sunlight, rainwater as well as the photocatalytic and superhydrophilic properties of TiO_2. Adsorbed organic materials such as oil can be decomposed by the photocatalytic property of TiO_2, whereas organic contaminants and dust can be washed off by rainwater because of the superhydrophilic property of TiO_2. It is not reasonable to assume that a superhydrophilic self–cleaning surface will never turn dirty. Note that self–cleaning processes are dependent on the illumination conditions, amount of rainfall and the accumulation rate of soilage, etc. Nonetheless, such a self–cleaning surface will retard the rate of contamination. This can save time and cost of cleaning maintenance of high buildings, sport arenas and flexible plastic materials (e.g. domes, canopies and tents) [42, 343].

TOTO Inc. is the main producer of TiO_2–based self–cleaning tiles in Japan. Its self–cleaning tiles have been used in more than 5000 buildings in Japan according to statistical data in 2003 [5]. TOTO has been collaborating with Professor Fujishima at University of Tokyo in the photo–catalyst technology. They succeeded in finding a novel phenomenon on the photocatalyst. When the surface of a photocatalytic film is exposed to light, the contact angle of the photocatalyst surface with water is reduced gradually. After enough exposure to light, the surface reaches super–hydrophilicity. In other words, it does not repel water at all, so the water cannot exist in the shape of a drop, but spreads flatly on the surface of the photocatalyst (Figure 18) [344].

Bacteria and viruses are decomposed on tile surfaces containing NS–TiO_2 due to the strong oxidizing properties of titanium dioxide. If the walls, ceiling and floor of operating room are covered with photocatalytic tiles, bacteria floating in the air may be killed as they come in contact with the surface (see Table 5).

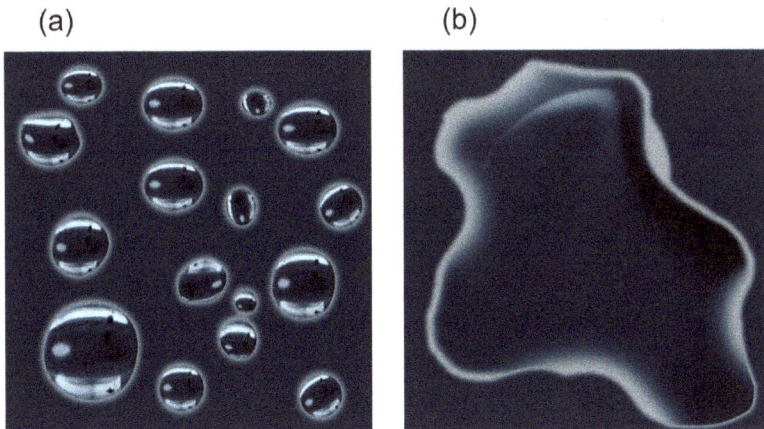

Figure 18. Comparing ordinary float glass (a) with anti–fogging glass with superhydrophilicity effect (b), were the water forms a thin sheet, with no fogging effect.

TiO_2 nanoparticles containing Ag^+ have been widely used as a filler in the manufacturing of antibacterial plastics, coatings, functional fibers, dishware and medical facilities. Ag^+ has a strong antibacterial activity even at lower concentrations. Commercial inorganic antibacterial TiO_2/Ag^+ nanoparticles (Shanghai Weilai Company, China), with primary particle size of about 70 nm and Ag^+ content of approximately 0.4% by weight were modified by grafting γ–aminopropyltriethoxysilane (APS). The modified particles were mixed in PVC to prepare composites whose antibacterial property was investigated. The results suggested that the modification had no negative effect on antibacterial activity of TiO_2/Ag^+ nanoparticles and PVC–TiO_2/Ag^+ composites exhibited a good antibacterial property [345].

TiO_2 and Fe^{3+}–doped coatings have been incorporated through sol–gel method on glass substrates. Trapalis *et al.* studied the antibacterial activity of TiO_2 (Fe^{3+}) nanostructured thin films against *E. coli*. The coatings exhibited a high antibacterial activity. This was enhanced with the increase of the temperature taking thermal treatment and the formation of anatase crystalline structures [346].

Sb–doped TiO_2 thin films were added to glass by RF magnetron co–sputtering method. The antibacterial activity against Escherichia coli bacteria was studied. The Sb–doped TiO_2 films showed high antibacterial activity eliminating *E. coli* bacteria [347].

NS–TiO_2–based self–cleaning glass is another important commercial self–cleaning product. Such glass can maintain extreme visual clarity even on a rainy day since water does not bead but instead spreads evenly across the surface (see Figure 18). If the amount of water is relatively small, the water layer becomes very thin and evaporates quickly. If the amount of water is significant, it forms a sheet–like layer that also has high visual clarity. It may be surprising that NS–TiO_2–coated glass can maintain the light transmittance properties of common glass; the higher refractive index of TiO_2 enhances the surface reflection. TiO_2 nanoparticles are dispersed in a SiO_2 matrix in the coating for self–cleaning glass. The composition of the coating is carefully controlled so that its refractive index is close to that of glass [5].

Table 5. Antibacterial and self–cleaning applications of *NS–TiO$_2$* [4, 5, 343–350].

Category	Function	Application
Road construction materials	Cleaning easiness, Self–cleaning, Anti–fogging	Tunnel lighting, Tunnel wall, Clear soundproof wall, Traffic signs, Lightning, Soundproofed wall, Guardrail, Road mirror, decorative laminated panel and Reflector on a read
Materials for residential and office buildings	Self–cleaning, Antibacterial, Anti–fogging, Accelerated drying	Parts of a Kitchen, a Bathroom and Interior furnishings, Exterior tiles, Siding boards, Window, Sash, Screen door, Gate door, Roof, Sun parlour, Mirror of a Bathroom and a Dresser, Toilet, Plastic surfaces, Aluminium siding, Building stone and curtains, Paper window blinds, Crystallized glass and Glass film
Materials for air cleaning	Self–Cleaning, Antibacterial	Room air cleaner, Photocatalyst–equipped air conditioners and interior air cleaner for factories, Outdoor air purifiers Concrete for highways, Roadways and footpaths, Tunnel walls, Soundproof walls and building walls
Materials for vehicles	Self–cleaning, Antibacterial, Anti–fogging, Preventing dewdrops forming	Painting and coating of vehicles, Outside of windows, Headlights, Inside of windows, Glass film, Helmet visor, Side–view mirror, Rear–view mirror and Windshield of a motorcycle, Side–mirror film
Materials for hospitals	Self–cleaning, Antibacterial	Cloth for hospital garments and uniforms, Tiles to cover the floor and walls of operating rooms, Silicone rubber for medical and hospital catheters
Self–cleaning glass	Self–cleaning, Antibacterial, Anti–fogging	Mirrors and glasses of vehicles, Windows, Conservatories, Glazed walls, Glass and conservatory roofs, Photocatalytic removal of pollutants (Stearic acid, Acid Blue 9 and Reactive Black 5)
Self–cleaning textiles	Self–cleaning, Antibacterial	Photodegradation of micelles, oils, solvents, sooth, aromatic, aliphatic hydrocarbons and stains of wine, coffee, make up and grease under daylight; self–cleaning tents

TiO_2 and Fe^{3+}–doped coatings have been incorporated through sol–gel method on glass substrates. Trapalis *et al.* studied the antibacterial activity of TiO_2 (Fe^{3+}) nanostructured thin films against *E. coli*. The coatings exhibited a high antibacterial activity. This was enhanced with the increase of the temperature taking thermal treatment and the formation of anatase crystalline structures [346].

Sb–doped TiO_2 thin films were added to glass by RF magnetron co–sputtering method. The antibacterial activity against Escherichia coli bacteria was studied. The Sb–doped TiO_2 films showed high antibacterial activity eliminating *E. coli* bacteria [347].

NS–TiO_2–based self–cleaning glass is another important commercial self–cleaning product. Such glass can maintain extreme visual clarity even on a rainy day since water does not bead but instead spreads evenly across the surface (see Figure 18). If the amount of water is relatively small, the water layer becomes very thin and evaporates quickly. If the amount of water is significant, it forms a sheet–like layer that also has high visual clarity. It may be surprising that NS–TiO_2–coated glass can maintain the light transmittance properties of common glass; the higher refractive index of TiO_2 enhances the surface reflection. TiO_2 nanoparticles are dispersed in a SiO_2 matrix in the coating for self–cleaning glass. The composition of the coating is carefully controlled so that its refractive index is close to that of glass [5].

Pilkington Activ ™ claims to be the original manufacturer of the world's first self–cleaning glass. Pilkington Activ ™ glass has a coating of nanocrystalline titanium dioxide that acts as an "invisible washer" to allow the surface to clean itself. The nano–scale (15 nm) coating has two remarkable properties that create the effect. It absorbs sunlight (ultraviolet radiation) creating a photocatalytic effect on dirt accumulated on the window and thus eliminating it. It also makes the surface of the glass hydrophyllic, that is, water forms a thin film that allows organic dirt particles to flow off and water to dry much faster. Pilkington Activ™ glass reduces the amount of sunlight passing through by only about 5%; the coating is tough enough and would require real abrasion like scrubbing with steel wool to take it off [348, 349]. Pilkington Activ™ self–cleaning glass has been investigated by Mills *et al.* [348] for oxidation of stearic acid films (see Table 5). Mills and co–workers found that for thick layers of stearic acid, the initial kinetics were zero–order for disappearance of the integrated area of CH_2– and CH_3– group IR bands of stearic acid. In the absence of stearic acid, contact angle measurements on Activ™ glass showed that 45 min of UV light exposure decreased the water contact angle from 67° (hydrophobic) to 8° (hydrophilic). Complete recovery of the water contact angle after UV illumination required 120–150 h.

Chin and Ollis [349] characterized the photocatalytic degradation properties of Activ™ glass using organic dyes (Acid Blue 9 and Reactive Black 5) deposited in a liquid film and dried on the Activ™ glass. The glass was oxidized to show visual decolorization and recovery of the aesthetic clarity of the glass and to determine the light–driven reaction kinetics on the Activ™ (see Table 5). Characterization of the Activ™ glass using atomic force microscopy yielded a *TiO$_2$* average particle diameter of 95 nm and an average thickness of 12 nm. X–ray diffraction results revealed only *TiO$_2$* in anatase phase. A *TiO$_2$* crystallite thickness of 18 nm was estimated using the Scherrer equation.

In addition to the above–mentioned applications, the super–hydrophilicity of *NS–TiO$_2$*–based glass can be applied for many other products utilizing secondary properties. For example, the super–hydrophilic surface dries quickly utilizing this property. This effect can be applied to prevent dewdrops forming inside a windowpane and a greenhouse, for the purpose of protecting vegetables from rotting by dewdrops.

Another important *NS–TiO$_2$*–based self–cleaning product is self–cleaning textiles and polymer film materials (see Table 5) [5, 350]. The tent material, made from flexible PVC film, is difficult to clean. Coating the PVC film with a *TiO$_2$* layer can resolve the soilage problem. However, an intermediate layer with an inorganic–organic graded structure is necessary to avoid the photodegradation of the PVC film by the *TiO$_2$* coating.

The self–cleaning effect of *NS–TiO$_2$* clusters on several textiles (e.g. bleached cotton and of mercerized cotton) under daylight irradiation has been reported [350]. The textile pre–treatment was carried out by RF–plasma, MW–plasma or vacuum–UV irradiation. The photo–discoloration and mineralization of stains of red wine, coffee, make up and grease in the presence of textiles containing *NS–TiO$_2$* were monitored upon daylight irradiation. One of the reasons to undertake that research was to explore the commercial application for self–cleaning purposes of *NS–TiO$_2$* semiconductor on cotton textiles prepared at relatively low temperatures. The use of *NS–TiO$_2$* loaded flexible substrates can possibly allow their application during the photodegradation of micelles, oils, solvents, sooth, aromatic and aliphatic hydrocarbons under daylight (see Table 5).

4.8. Electrocatalysis

Electrochemical oxidation of methanol has been studied extensively in conjunction with the development of fuel cells. Methanol is catalytically oxidized on platinum electrodes producing CO_2 and six electrons per CH_3OH molecule, making this organic compound a promising candidate for direct oxidation fuel cells. Unfortunately, if not prevented by some physical or chemical conditions a catalytic poison namely methanolic CO develops impeding the oxidation. Basic research aimed at optimizing the methanol oxidation rates via surface structural adjustment using well–defined, single–crystal electrodes (see Table 4) [351–354]. Recently, Hepel *et al.* [352] reported the development of a new nanostructure catalyst material for methanol oxidation reaction based on anodic nanoporous TiO_2 films. Their new improved TiO_2 films have been grown to form regular cylindrical nanopores (nanotubes) with a diameter range of 20–80 nm. The PtRu cluster catalysts and PtFe nanoparticles were deposited on the nanoporous TiO_2 matrix and activated using the point defect formation protocols. This results in TiO_2 supported bimetallic catalysts (TiO_2|PtRu and TiO_2|$Pt_{0.3}Fe_{0.7}$). This bimetallic catalyst was used in methanol oxidation.

Macak *et al.* [353] demonstrated that nanotubular TiO_2 matrices (generated from mixed H_2SO_4/HF electrolytes using the self–organizing anodization process) may act as active support for catalytic Pt/Ru nanoparticles during electrooxidation of methanol. The nanotubular TiO_2 layers consist of individual tubes of 100 nm diameter, 500 nm length and 15 nm wall thickness. They found that nanotubular TiO_2 support provided a high surface area and it significantly enhanced the electrocatalytic activity of Pt/Ru for methanol oxidation.

In another work, Xia *et al.* [354] studied the electrocatalytic properties of PtRu/C–TiO_2 toward the oxidation of methanol. PtRu/C–TiO_2 catalysts were prepared by sol–gel method using titanium precursor $(Ti(OBu)_4)$ on PtRu/C. They described that the activities of the catalysts toward methanol oxidation decreased in the following order. PtRu/C–TiO_2 after sintering > PtRu/C–TiO_2 > PtRu/C > PtRu/C after sintering.

Marken *et al.* [355] reported electrocatalytic oxidation of nitric oxide in TiO_2–Au nanocomposite film electrodes. In this work, TiO_2 (anatase) nanoparticles (6 nm diameter) and gold nanoparticles (20 nm diameter) are formed via a layer–by–layer deposition procedure. TiO_2 nanoparticles were deposited with a Nafion polyelectrolyte binder followed by calcination to give a mesoporous thin film electrode. Gold nanoparticles were incorporated into this film employing a polyelectrolyte binder followed by calcination to give a stable mesoporous TiO_2–gold nanocomposite. Electrochemical experiments have been performed in aqueous KCl, buffer solutions, nitric oxide (NO) and nitrite in phosphate buffer solution. It has been shown that the NO oxidation occurred as a highly effective electrocatalytically amplified process at the surface of the gold nanocomposite with the co–evolution of oxygen, O_2. In contrast, the oxidation of nitrite to nitrate occurred at the same potential but without oxygen evolution.

Feng *et al.* [356] developed a high performance electrochemical wastewater treatment system using pulse voltage and evaluated its performance using domestic wastewater, pond water containing algae and wastewater from hog raising. In that study, the cathode was made from a titanium sheet and the anodes were a sheet of titanium, platinum and titanium coated with Ti/RuO_2–TiO_2. The ratio of RuO_2 to TiO_2 was 30:70 (v/v). The titanium supported oxide layer was coated by thermal decomposition of precursors in isopropyl alcohol.

4.9. Photocatalytic Applications of Titanium Dioxide Nanomaterials

4.9.1. Pure titanium dioxide nanomaterials

In this section, the application of NS–TiO_2 materials in the field of photocatalysis is considered. Although the application of NS–TiO_2 materials is mentioned in other chapters, in this section we describe extensively.

Since the Honda–Fujishima effect [5, 206] was reported in the early 1970s, extensive studies of photocatalysis on semiconductors, in particular on illuminated surfaces of titanium dioxide, have been carried out [6–19]. Through the 1970s and 1980s the main interest was focused on hydrogen photoevolution from water or organic waste. At this point, the properties of semiconductors were extensively investigated and described leading to semiconductor modifications, sensitization and improvement in hydrogen evolution. Although water splitting is not in practical use yet, some progress was made. In the 1990s, the topic shifted to the applications for environmental remediation using TiO_2 as a photocatalysts and some progress has taken place [27].

Photocatalytic chemistry involving semiconductor materials has grown from a subject of esoteric interest to one of central importance in both academic and technological research. In this context, environmental pollution and its control through nontoxic treatments and easy recovery processes is a serious matter. The number of publications concerning mineralization of dyes, pesticides, fungicides and hazardous compounds, etc., increased tremendously in the last decade [16, 17, 20, 27, 357].

Photocatalysis covers the range of reactions proceeding under the action of light. Among these, we find catalysis of photochemical reactions, photo–activation of catalysts and photochemical activation of catalytic processes. Photocatalysis is defined by the IUPAC. "Photocatalysis is the catalytic reaction involving light absorption by a catalyst or a substrate" [358, 359]. A more precise definition may be that "Photocatalysis is a change in the rate of chemical reactions or their generating under the action of light in the presence of the substances (photocatalysts) that absorb light quanta and are involved in the chemical transformations of the reaction participants, repeatedly coming with them into intermediate interactions and regenerating their chemical composition after each cycle of such interactions" [359].

The most typical processes covered by photocatalysis are the photocatalytic oxidation (PCO) and the photocatalytic decomposition (PCD) of substrates such as organic compounds. The PCO process employs the use of gas–phase oxygen as a direct participant to the reaction, while the PCD takes place in the absence of O_2 [55].

Several semiconductors possess band gaps suitable to catalyze chemical reactions. Titanium dioxide has become a "gold standard" semiconductor in the field of photocatalysis. TiO_2 is chemically and biologically inert as well as cheap to manufacture and apply. In recent years, applications of $NS–TiO_2$ in environmental remediation have been one of the most active areas in research [30].

Several researchers focused on TiO_2 nanoparticles and its application as a photocatalyst in water treatment. Nanoparticles that are activated by light, such as the large band–gap semiconductors titanium dioxide and zinc oxide, are frequently studied for their ability to remove organic contaminants from various media. These particles have the advantages of being readily available, inexpensive and have low toxicity. The semiconducting property of TiO_2 is necessary for the removal of organic pollutants through the excitation of TiO_2 semiconductor with an energy source greater than its band gap to generate electron hole pairs (see Figure 5). This characteristic may be exploited in different reduction processes at the semiconductor/solution interface. Although the exact mechanism differs from one pollutant to the next, it has been widely recognized that superoxide and specifically hydroxyl radicals act as active species in the degradation of organic compounds [46, 66].

A semiconductor can be doped with donor atoms to provide electrons to the conduction band. Semiconducting materials can also be doped with acceptor atoms that take electrons from their valence band and leave behind some positive charges (holes). The most effective properties of semiconducting nanoparticles are noticeable changes in their optical properties which differ from their bulk counterpart materials. There is a significant shift in the optical absorption spectra toward the blue region (shorter wavelengths) as the particle size is reduced [360].

Stathatos *et al.* [361] used a reverse micelle technique to make TiO_2 nanoparticles and deposit them as thin films. The research group deposited TiO_2 mesoporous films on glass slides by dip–coating in reverse micellar gels containing titanium isopropoxide. The films exhibited a high capacity for adsorption of several dyes from aqueous or alcoholic solutions. When the colored films were exposed to visible light, they carried out a rapid degradation of the adsorbed dyes.

The semiconducting properties of TiO_2 materials are great in general. However, the rapid recombination of photo–generated electron hole pairs resulting from small charge separation distances within the particle and the non–selectivity of the system may present limitation in the application of $NS–TiO_2$ in photocatalysis processes. To avoid the rapid recombination of electron holes, charge separation gaps can be increased by introducing a deeper trapping site outside the semiconductor particle. It was suggested that interfacial electron transfer could take place using surface Ti (IV) atoms due to their coordination with solvent molecules. This can generate constitute trapping sites for the conduction band electrons while hole transfer occurs through surface oxygens [362].

Replacing adsorbed solvent molecules and ions by chelating agents, a method known as surface modification, changes the energetic situation of such surface states and may considerably change the chemistry taking place at the surface of titanium dioxide [362].

The effect of surface modification of nanocrystal TiO_2 with specific chelating agents such as arginine, lauryl sulfate and salicylic acid was investigated by the photocatalytic degradation of nitrobenzene (NB). The results of the study are shown in Table 6 [362].

Phenol is one of the toxic materials in municipal and wastewater. Titanium dioxide nanoparticles of both anatase and rutile forms were synthesized by hydrothermal treatment of microemulsions and used in the wet oxidation of phenol [363]. The advantage of this method of preparation is that the size of particles can be affected by the ratio of surfactant to water. Size of water droplets in the reverse microemulsions is approximately the same as that of formed particles. The main reactions in phenol degradation are [363]:

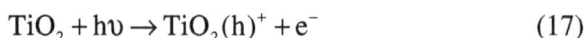

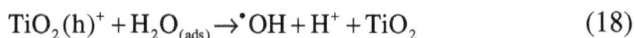

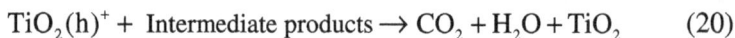

$$TiO_2 + h\upsilon \rightarrow TiO_2(h)^+ + e^- \tag{17}$$

$$TiO_2(h)^+ + H_2O_{(ads)} \rightarrow {}^{\bullet}OH + H^+ + TiO_2 \tag{18}$$

$${}^{\bullet}OH + \text{Phenol} \rightarrow \text{Intermediate products (e.g., benzoquinone)} \tag{19}$$

$$TiO_2(h)^+ + \text{Intermediate products} \rightarrow CO_2 + H_2O + TiO_2 \tag{20}$$

A novel composite reactor through combination of photochemical and electrochemical systems was used for the degradation of organic pollutants [364]. In this process UV excited nanostructure TiO_2 were used as the photocatalyst. The reactor was evaluated by the degradation process of Rhodamine 6G (R–6G).

Fine TiO_2 nanoparticles are more efficient than the immobilized catalysts in the degradation of organic compounds found in wastewater. However, complete separation and recycling of these fine TiO_2 particles (less than 0.5 µm) from the treated water is very expensive. Thus, the method from an economic point of view is not suitable for industrial applications. This problem can be solved by fixing TiO_2 nanoparticles on supports such as glass plates, aluminium sheets and activated carbon. The photocatalytic activity of carbon–black–modified nano–TiO_2 (CB–TiO_2) thin films has shown at least 1.5 times greater than that of TiO_2 thin films in the degradation of reactive Brilliant Red X–3B [365].

Core SrFe$_{12}$O$_{19}$ nanoparticles–*TiO$_2$* nanocrystals were also synthesized as magnetic photocatalytic particles [366]. In this case the photocatalyst particles are recovered from the treated water stream by applying an external magnetic field (see Table 6).

Many problems may occur when natural organic matters are present in the water, since they can occupy the catalyst active sites causing much lower decomposition efficiency. A combination of adsorption and oxidative destruction techniques may become a useful method to overcome the above problem. Ilisz *et al.* [367] used a combination of *TiO$_2$*–based photocatalysis and adsorption processes to test the decomposition of 2–chlorophenol (2–CP). The group created three systems which are presented below:

(1) *TiO$_2$* intercalated into the interlamellar space of a hydrophilic montmorillonite by means of a heterocoagulation method (*TiO$_2$* pillared montmorillonite, TPM);

(2) *TiO$_2$* hydrothermally crystallized on hexadecylpyridinium chloride–treated montmorillonite (HDPM–T);

(3) Hexadecylpyridinium chloride–treated montmorillonite (HDPM) applied as an adsorbent and Degussa P25 *TiO$_2$* as a photocatalyst (HDPM/*TiO$_2$*) [9].

The latter showed the highest rate for the pollutant decomposition compared to the others. The study revealed that the system could be re-used without further regeneration.

In another application, the work was focused on crystalline titania with ordered nanodimensional porous structures [368]. In this case, the mesoporous spherical aggregates of anatase nanocrystal were fabricated and cetyltrimethylammonium bromide (CTAB) was employed as the structure–directing agent. The interaction between cyclohexane micro-droplets and cetyltrimethylammonium bromide self–assemblies was applied to photodegrade a variety of organic dye pollutants in aqueous media such as methyl orange (see Table 6).

Also, in the study of Peng *et al.* [369], the mesoporous titanium dioxide nanosized powder was synthesized using hydrothermal process by applying cetyltrimethylammonium bromide as surfactant–directing and pore–forming agent. They synthesized and applied this nanoparticle for the oxidation of Rhodamine B (see Table 6).

Table 6. Removal of pollutants using TiO₂ nanoparticles.

Type of nanoparticle	Removal target	Initial concentration	Dose of nanoparticle	Irradiation time (min)	Removal efficiency (%)	Ref.
80% anatase and 20% rutile (Degussa P25)	Basic Red 46	17.5 mg/L	Immobilized on glass beads	90	80	[41]
TiO_2 nanoparticle	Acid red 14	20 mg/L	40 mg/L	210	100	[46]
Degussa P25	Direct Red 23	10 mg/L	Immobilized on glass beads	180	80	[56]
Degussa P25	Herbicide, Erioglaucine	20 mg/L	150 mg/L	25	100	[66]
Synthesized rutile–anatase	Acid blue 9	20 mg/L	150 mg/L	90	100	[357]
Arginine–modified TiO_2	Nitrobenzene	50 mg/L	$[TiO_2]=0.1$ mol/L $[SM]^1=0.03$ mol/L	120	100	[362]
Anatase TiO_2	Phenol	–	1.8 g/L	408	100	[363]
TiO_2 nanoparticle	Rhodamine 6G	125 mmol/L	0.1 %(w/w)	12	90	[364]
$TiO_2/SrFe_{12}O_{19}$ composite	Procion Red MX–5B	10 mg/L	(2.0 mg) 30% TiO_2	300	98	[366]
Mesoporous Anantase nanocrystal	Methyl orange	30 mg/L	3 g/L	45	100	[368]

Table 6 (continued).

Type of nanoparticle	Removal target	Initial concentration	Dose of nanoparticle	Irradiation time (min)	Removal efficiency (%)	Ref.
Mesoporous TiO₂ nanopowder[b]	Rhodamine B	1.0×10^{-5} mol/L	50 mg/50 mL	120	97	[369]
Mesoporous titania nanohybrid (naohybrid–I)[c]	4-chlorophenol	1.0×10^{-5} mol/L	25 mg/100 mL	240	99	[370]
Mesoporous titania nanohybrid (naohybrid–I)[c]	Methyl orange	1.0×10^{-5} mol/L	25 mg/100 mL	120	100	[370]
Rutile TiO₂ nanoparticle	Parathion	50 mg/L	1000 mg/L	120	>70	[373]
TiO₂/AC nanoparticle[d]	Methyl orange	1.0×10^{-3} mol/L	0.5 g/200 mL (47 wt% TiO₂)	100	77	[374]
TiO₂/AC nanoparticle[d]	Methyl orange	1.0×10^{-3} mol/L	0.5 g/200 mL (63wt% TiO₂)	100	66	[374]
TiO₂ nanoparticle	Basic dye	20 mg/L	1.22 g/L	180	>80	[375]

[a] Surface modifier
[b] Calcinated at 400°C
[c] [Ti] nanoparticles/[Ti] layered titanate
[d] TiO₂ + activated carbon
 T=25°C, 150 rpm

The mesoporous structure, with high surface area could provide simple accessibility of guest molecules to the active sites and increase their chances to receive light. One research group fabricated mesoporous photocatalysts with delaminated structure. The exfoliated layered titanate in aqueous solution was reassembled in the presence of anatase TiO_2 nanosol particles to make a great number of mesopores and increase the surface area of TiO_2 [370] (see Table 6).

Degussa P25 TiO_2 is a highly photoactive form of TiO_2 composed of 20–30% rutile and 70–80% anatase TiO_2 with particle size in the range of 12 to 30 nm. Adams *et al.* [371] synthesized SBA–15 mesoporous silica thin films encapsulating Degussa P25 TiO_2 particles via a block copolymer templating process. High calcination temperatures (above 450°C) are typically required to form a regular crystal structure. However, heat treatment at high temperature, can decline the surface area and loose some surface hydroxyl or alkoxide group on the surface of TiO_2. This problem was solved by hydrothermal process to produce pure anatase– TiO_2 nanoparticles at low temperature (200° C, 2 h). These TiO_2 nanoparticles have several advantages, such as fully pure anatase crystalline form, fine particle size (8 nm) with more uniform distribution and high–dispersion either in polar or non–polar solvents, stronger interfacial adsorption and convenient coating on different supporting materials.

The behavior of anatase nano–TiO_2 in catalytic decomposition of Rhodamine B dye was also examined [372]. Rhodamine B was fully decomposed with the catalytic action of nano–TiO_2 in a short time (i.e. 60 min). Photocatalytic activity of the nano–TiO_2 for degradation of RB was compared with Degussa P25 at optimum catalysis conditions determined for the nano–TiO_2. Repeatedly usage of the synthesized catalyst was compared with Degussa P25 and it was found that the nano–TiO_2 showed higher photocatalytic activity than Degussa P25, even after the fourth use.

In recent years, the technology of ultrasonic degradation has been studied and extensively used to treat some organic pollutants. The ultrasound with low power was employed as an irradiation source to make heat–treated TiO_2 powder. This method was used for decomposition of parathion with the nanometer rutile titanium dioxide (TiO_2) powder as the sonocatalyst after treatment in high–temperature activation [373].

An appropriate method for increasing the photocatalytic efficiency of TiO_2 consists in adding a co–adsorbent such as activated carbon (AC). This synergy effect has been explained by the formation of a common contact interface between different solid phases. Activated carbon acts as an adsorption trap for the organic pollutant which is then efficiently transferred to the TiO_2 surface where it is immediately degraded by a mass transfer to the photoactivated TiO_2. For this reason, carbon grain coated with activated nano–TiO_2 (20–40 nm) (TiO_2/AC) was prepared and used for the photodegradation of methyl orange (MO) dyestuff in aqueous solution (see Table 1) [365]. Some of the benefits that took place in the application of these activated carbons are summarized below [44, 374]:

(1) The adsorbent support makes a high concentration environment to target organic substances around the loaded TiO_2 particles by adsorption. Therefore, the rate of photooxidation is enhanced.

(2) The organic substances are oxidized on the photocatalyst surfaces via adsorption states. The resultant intermediates are also adsorbed and then further oxidized. Toxic intermediates, if formed, are not released in the air and/or in solution thus preventing secondary pollution.

(3) Since the adsorbed substances on the adsorbent supports are finally oxidized to give CO_2, the high adsorbed ability of the hybrid photocatalysts for organic substances is maintained for a long time. The amount of TiO_2 as catalyst may play a significant role upon the photo–efficiency of hybrid catalysts.

Wu *et al.* [375] studied dye decomposition kinetics in a batch photocatalytic reactor under various operational conditions including agitation speed, TiO_2 suspension concentration, initial dye concentration, temperature and UV illumination intensity in order to establish reaction kinetic models (see Table 6).

Photocatalytic removal of Acid Red 14 (AR14), commonly used as a textile dye, using TiO_2 suspensions irradiated by a 30W UVC lamp has been studied [46, 376]. It was found that TiO_2 and UV light had a negligible effect when they were used on their own. The mechanism of photocatalysis is described in Figure 5. In this study, the effects of some parameters such as pH, the amount of TiO_2 and initial dye concentration have also been examined. The photodegradation of Acid Red 14 was enhanced by the addition of proper amount of hydrogen peroxide, but it was inhibited by ethanol. From the inhibitive effect of ethanol it was concluded that hydroxyl radicals play a significant role in the photodegradation of dye whereas a direct oxidation by positive holes was probably not negligible (see Table 6).

Due to the intensive agricultural methods implemented around the global, in the last few decades the variety and quantities of agrochemicals present in continental and marine natural waters have dramatically increased. Photocatalytic degradation of Erioglaucine as an herbicide in the presence of P25 TiO_2 nanoparticles under UV light illumination has been reported [66]. The photocatalytic activities between the commercial TiO_2 Degussa P25 and a rutile TiO_2 was compared. It was found that the higher photoactivity of TiO_2 P25 as compared to that of rutile TiO_2 in the photodegradation of erioglaucine may be due to higher hydroxyl content, higher surface area, nano–size and crystallinity of the Degussa P25 (anatase–rutile). The influence of basic photocatalytic parameters such as pH of the solution, initial concentration of erioglaucine, amount of TiO_2 and irradiation time on the photodegradation efficiency was also reported. Experimental results indicated that the photocatalytic degradation process could be explained in terms of the Langmuir–Hinshelwood kinetic model [66].

In general, it can be concluded that all modified and thin film samples prevent rapid recombination, while CB–TiO_2 films and TiO_2/strontium ferrite samples have the advantage of easy separation because of their fixation on the support.

It has long been observed that mixed–phase preparations of TiO_2 containing both anatase and rutile tend to exhibit higher photocatalytic activities than pure–phase TiO_2. The best–known example of this phenomenon is Degussa P25, which consists of about 70–80% anatase and the remainder rutile, with traces of brookite and amorphous phases [26, 30]. Furthermore, when Bacsa and Kiwi [377] prepared TiO_2 samples with a range of anatase:rutile ratios; the highest photoactivity was obtained with a 70:30 ratio, which is similar to that of P25.

Possible reasons for the improved performance of such TiO_2 samples have been unclear. It was hypothesized that rutile acts as a sink for the electrons generated in anatase allowing physical separate the electron and hole and thereby depressing rates of recombination [26, 378] (see Figure 19 a). This model is consistent with the fact that the band edges of rutile lie within those of anatase; i.e., the potential of the conduction band edge of anatase is more negative than that of rutile. However, it has been recently shown that just the opposite occurs. Rutile undergoes band gap activation and electrons are shuttled from rutile to anatase sites which must be of lower energy (see Figure 19 b). This implies that one or more trap sites exist on anatase at potentials more positive than the conduction band edge of either anatase or rutile. This was recently confirmed by a photoacoustic spectroscopy study of anatase, which found trap sites on anatase have an average of 0.8 eV below the conduction band edge [379].

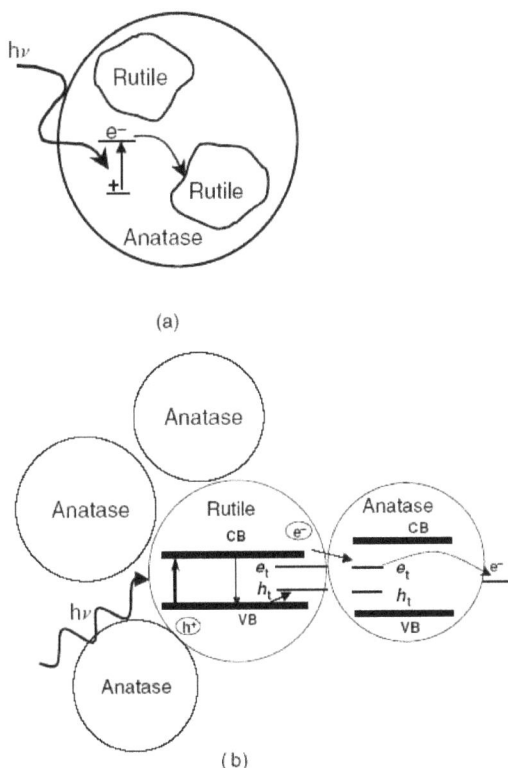

Figure 19. Models of Degussa P25: (a) conventional wisdom holds that rutile islands surround anatase particles and rutile is an electron sink; (b) new picture involves a small rutile core surrounded by anatase crystallites, where electrons are shuttled from rutile to anatase. (Adapted from Khataee *et al.* [357] with permission from publisher, Taylor & Francis. License Number: 2627090166720).

It can be concluded that the size and morphology of rutile and anatase nanocrystals are critical to the separation and enhanced activity of mixed–phase catalysts like Degussa P25. As it was illustrated in Figure 19 b, an emerging model of P25 particles describes a typically small rutile core surrounded by anatase crystallites. Catalytic "hot spots" are believed to exist at the intersection of the two phases, where distorted geometry gives rise to unique surface chemistry [380]. It has been confirmed that most recombination in the mixed–phase of P25 occurs not within the lattice but at surface sites on both anatase and rutile phases [381–384].

4.9.2. TiO$_2$–Based Nanoclays

Clays are layered minerals with space between the layers where they can adsorb positive and negative ions as well as water molecules. Clays undergo exchange interactions of adsorbed ions with the outside too. Although clays are very useful for many applications, they have one main disadvantage i.e. lack of permanent porosity. To overcome this problem, researchers are looking for a way to prop and support the clay layers with molecular pillars. Most of the clays can swell and thus increase the space between their layers to accommodate the adsorbed water and ionic species. These clays are employed in the pillaring process.

As expressed previously, ultra fine TiO_2 powders have large specific surface area, but due to their easily agglomeration into larger particles, an adverse effect on their catalytic performance has been observed. It has been shown that the recovery of pure TiO_2 powders from water was very hard when they were used in aqueous systems. Dispersing TiO_2 particles in layered clays appear to provide a feasible solution to such problems. Such composite structures, known as pillared clay, can stabilize TiO_2 particles and maintain the surface of TiO_2 crystals to access different molecules. In addition, the interlayer surface of pillared clays is generally hydrophobic, which is an advantage in the adsorption and enrichment of diluted hydrophobic organic compound in water (see Table 7) [385].

Ooka *et al.* [386] prepared four kinds of *TiO$_2$* pillared clays with montmorillonite, saponite, fluorine hectorite and fluorine mica. The group presented the surface hydrophobicities and performances of these clays in adsorption– photocatalytic decomposition of phthalate esters. They found out that surface hydrophobicity of pillared clays largely varied with the host clay. Nonetheless, employing the host clays can improve the surface hydrophobicity of *TiO$_2$* pillared clays. The *TiO$_2$* particles in the pillared clays are too small to form a crystal phase. Therefore, they exhibit a poor photocatalytic activity. To overcome this problem, nanocomposites of titanium dioxide and silicate nanoparticles were made by a reaction between titanium hydrate sol of strong acidity and smectite clays in the presence of polyethylene oxide (PEO) surfactants [19]. As a result, larger precursors of *TiO$_2$* nanoparticles formed and condensed on the fragmentized pieces of the silicate. Introducing PEO surfactants into the synthesis process can significantly enhance the porosity and surface area of the composite solid [386].

Choy *et al.* [387] prepared highly porous layered inorganic–inorganic nanohybrids (surface area ~590 m^2/g) by pillaring SiO$_2$–*TiO$_2$* nanosol particles with aluminosilicate layers. The sorption behavior of various solvent vapors such as hexane, methanol and water revealed internal pore surfaces of SiO$_2$–*TiO$_2$* pillared aluminosilicate (STPC) to be hydrophobic. From a distinct blue shift of absorption edge in UV–vis spectra researchers found that the nanosized *TiO$_2$* particles are formed between silicate layers as a pillar. They also indicated that the pillared titania existed in the form of anatase–structured nanocrystals, not in the form of covalently bonded mixed particles of *TiO$_2$*–SiO$_2$. On the basis of their findings, it can be concluded that the quantum–sized *TiO$_2$* and SiO$_2$ particles are independently intercalated to form a multilayer stacking intracrystalline structure in the gallery space of aluminosilicate clay.

Nanocomposites of iron oxide and silicate were also synthesized for the degradation of azo–dye Orange II (see Table 7) [388, 389]. To improve the sorption capacity, clays were modified in different ways (e.g. treatment by inorganic and organic compounds). Organoclays have recently attracted lots of attention in a number of applications, for example, dithiocarbamate–anchored polymer/organosmectite for the removal of heavy metal ions from aqueous media (see Table 7) [390].

A new class of nano–sized large porous titanium silicate (ETAS–10®) and aluminum–substituted ETAS–10 with different Al_2O_3/TiO_2 ratio were successfully synthesized and applied to remove heavy metals in particular Pb^{2+} and Cd^{2+} (see Table 7). Since tetravalent Ti is coordinated by octahedral structure, it creates two negative charges that must be normally balanced by two monovalent cations. This leads to a great interest in ion exchange or adsorption property of this material [391].

Zhu *et al.* [392] prepared thermally stable composite nanostructures of titanium dioxide (anatase) and silicate nanoparticles from Laponite clay and a sol of titanium hydrate in the presence of polyethylene oxide (PEO) surfactants. Laponite is a synthetic clay that readily disperses in water and exists as exfoliated silicate layers of about 1–nm thick in transparent dispersions at high pH values. The group found out that the composite solids exhibited superior properties as photocatalysts for the degradation of Rhodamine 6G in aqueous solution in comparison with TiO_2 P25 (see Table 7). The BET surface area of P25 is about 50 m^2/g, while the composites have surface areas of 300–600 m^2/g. With such large surface areas, the composite samples exhibited a binary function for removing organic compounds from water through both photocatalysis and adsorption. They also attributed the superior catalytic performance of the TiO_2 nanocomposites to the small size (3–9 nm) of the TiO_2 crystals in the samples compared to that in P25 (about 25 nm).

From these results and the list of the removed pollutants using nanoclays (Table 7), it is believed that TiO_2/clay composites are promising heterogeneous nanocatalysts for the photocatalytic removal of water contaminates.

Table 7. Removal of pollutants using nanoclays.

Type of nanoparticle	Removal target	Initial concentration	Dose of nanoparticle	Contact time (min)	Removal efficiency (%)	Adsorption capacity	Ref.
Fe–nanocomposite	Azo–dye orange II	0.1 mM	1.0g Fe nanocomposite/L + 4.8 mM H$_2$O$_2$+1×8 W UVC[b]	>20	>90	–	[388, 389]
Dithiocarbamate–anchored nanocomposite	Pb(II), Cd(II), Cr(III)	–	–	–	–	170.70, 82.20, 71.10 mg/g	[390]
ETAS–10 (A)[c]	Pb(II)	–	–	–	–	1.75 mmol/g	[391]
ETAS–10(A)	Cd(II)	–	–	–	–	1.24 mmol/g	[391]
ETAS–10(B)[d]	Pb(II)	–	–	–	–	1.68 mmol/g	[391]
ETAS–10(B)	Cd(II)	–	–	–	–	1.12 mmol/g	[391]
Composites of TiO$_2$ (Anatase) and silicate nanoparticles	Rhodamine 6G	2 × 10^{-5} M	25 g/L	60	>90	–	[392]
TiO$_2$/clay composites	Herbicide Dimethachlor	2 mg/L	200 mg/L	180	>90	2–17 %	[393]
Bentonite clay–based Fe nanocomposite	Orange II	0.2 mM	1.0 g/L and 10 mM H$_2$O$_2$	120	>99	–	[394]
TiO$_2$/ HDPM clay[e]	2–chlorophenol	5 g/L	1 g/L	1080	–	–	[395]

[a] Pseudo–second order parameters,
[b] UV irradiation,
[c] ETAS–10 (A) : (Al$_2$O$_3$/TiO$_2$=0.1), T=25°C,
[d] ETAS–10(B): (Al$_2$O$_3$/TiO$_2$=0.2), T=25°C,
[e] Hexadecylpyridinium chloride–modified montmorillonite.

4.9.3. Metal ions and non–metal atoms doped nanostructured TiO₂

One of the major challenges for the scientific community includes the proper application of NS–TiO_2 under visible light. Visible light composes the largest part of solar radiation. Anatase, which is the most photoactive phase of TiO_2, only absorbs ultraviolet light with wavelengths shorter than 380 nm. The content of ultraviolet light in indoor illumination is significantly smaller than that in sunlight, because the fluorescent lamp mainly emits visible light. To use solar radiation efficiently to conduct photocatalysis, two main approaches have been proposed:

(I) The first approach consists in doping a photocatalysts with transition metal ions (e.g. Cr^{3+}, Fe^{3+}) that can create local energy levels within the band gap of the photocatalyst. That corresponds to the absorption bands lying in the visible light spectrum. It was assumed that the photoexcitation of such impurities should lead to the generation of free charge carriers to initiate surface chemical processes. However, the efficiency of such systems under visible light strongly depended on the preparation method. In some cases, such doped photocatalysts showed no activity under visible light and lower activity in the UV spectral range compared to the non–doped photocatalyst. This was due to high carrier recombination rates through the metal ion levels. In addition, doped materials suffer from a thermal instability, an increase of carrier–recombination centers and required an expensive ion–implantation facility [5, 396].

(II) Another approach has been to dope TiO_2 with non–metal atoms, such as N, S, C and B. The mechanisms for both of these approaches are shown in Figure 20.

According to literature, the second approach tends to be better for the development of photocatalysts that use visible light. Commercial visible light–type TiO_2 photocatalysts are based on the anion–doped TiO_2. The anion doped NS–TiO_2 filters for air cleaners are available commercially [5, 396–402]. It is apparent that all of the applications of TiO_2 in photocatalysis discussed above can be promoted with visible light–type TiO_2, especially for indoor antibacterial and self–cleaning applications. The present problem for the anion–doped TiO_2 photocatalysts is that in some cases their photocatalytic activities under visible light are much lower than those under ultraviolet light. Much effort must be devoted to overcome this obstacle. So, we have considered the preparation and applications of non–metal doped TiO_2 nanomaterials below.

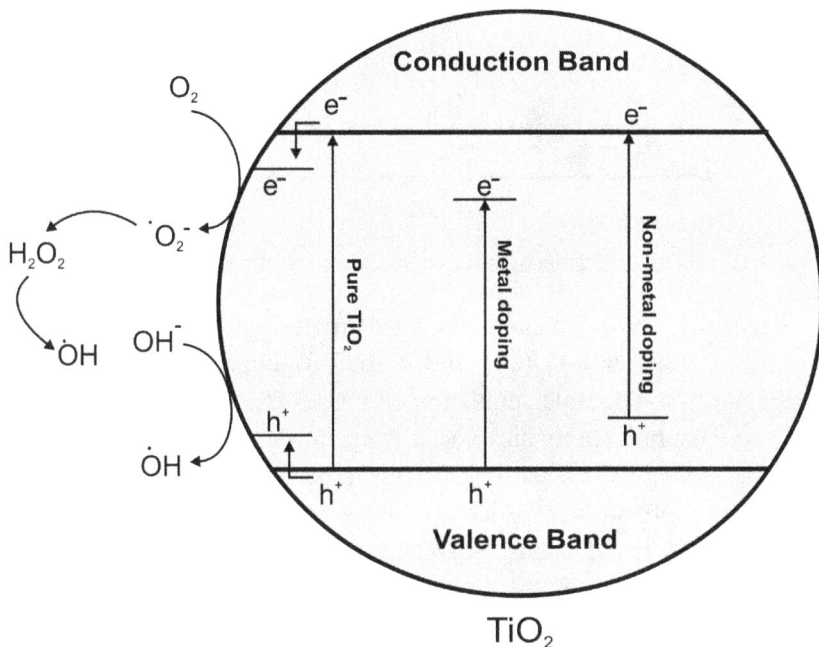

Figure 20. Mechanism of photocatalysis in the presence of pure, metal ions and non–metal atoms doped nanostructured TiO_2.

Asahi *et al.* [396] reported significant red shifts (up to 540 nm) of the spectral limit in photoactivity of TiO_2 doped with nitrogen (N). The group interpreted such results in terms of band gap narrowing due to mixing of the *p* states of the dopants with O 2*p* states forming the valence band of TiO_2, as illustrated in Figure 21. The researchers have calculated the densities of states for the substitutional doping of C, N, F, P, or S to replace O in the anatase crystal [397]. Among these atoms, it has been suggested that doping with N would prove to be most effective because its *p* states contribute to band–gap narrowing by mixing with the *p* states of oxygen. Structure of N–doped TiO_2 is shown in Figure 22.

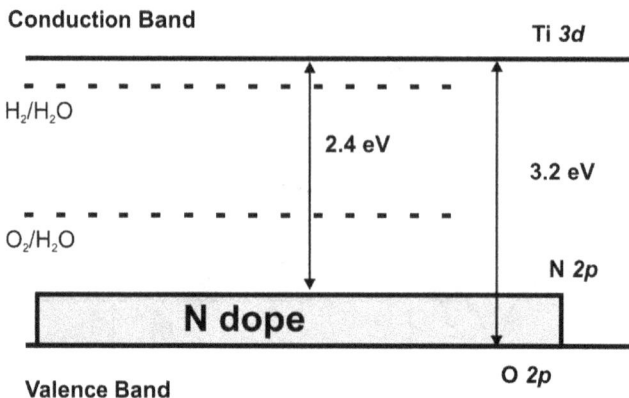

Figure 21. Schematic diagram of the electronic structures of pure and N–doped TiO_2.

Song *et al.* [398] calculated the band structures and charge densities of nitrogen (N)–, carbon (C)– and boron (B)–doped titanium dioxide through the first–principle simulation with CASTEP code. This is a state of the art quantum mechanics based program designed specifically for solid state materials science [399]. Researchers described that three $2p$ bands of impurity atom were located above the valence–band maximum and below the Ti $3d$ bands. Also, that along with the decreasing of impurity atomic number the fluctuations became more intensive. The group could not observe obvious band–gap narrowing in their results. Therefore, the cause of absorption in visible light might be the isolated impurity atom $2p$ states in band–gap rather than the band–gap narrowing. Irie *et al.* [400] also suggested that the visible–light response in N–doped TiO_2 might be due to N $2p$ states isolated above the valence–band maximum of TiO_2. Similarly, the red–shift in C–doped TiO_2 was observed by Choi *et al.* [401]. Moon *et al.* [402] reported the absorption band shifted towards longer wavelengths in B/TiO_2.

Various non–metal elements such as B, C, N, F, S, V and Br have been successfully doped into TiO_2 nanomaterials. Tables 8 and 9 provide more details.

N–doped TiO_2 was prepared by a wet method, i.e., the hydrolysis of titanium tetra–isopropoxide (TTIP) or titanium tetrachloride with an aqueous ammonia solution, followed by calcination at temperatures above 330°C [403]. The maximum absorption of visible light by this N–doped TiO_2 was about 50% at around 440 nm. To evaluate the photocatalytic activity of samples, the photooxidation of CO was carried out under irradiation of UV (365 nm) and visible (433 nm) monochromatic light (see Table 8).

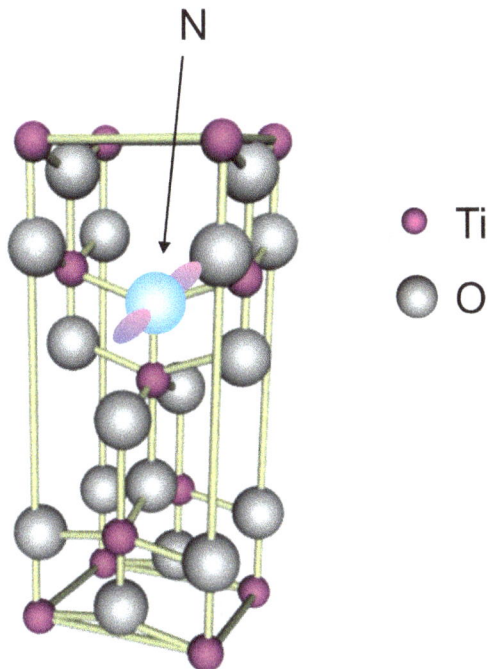

Figure 22. Unit cell of N–doped anatase TiO_2.

Nosaka *et al.* [404] reported doping of nitrogen atoms in commercially available TiO_2 powders by using organic compounds such as urea and guanidine. Figure 23 shows a flow chart for the preparation of N–doped TiO_2 nanopowder. A significant shift of the absorption edge to a lower energy and a higher absorption in the visible light region were reported. These N–doped TiO_2 powders exhibited photocatalytic activity for the decomposition of 2–propanol in aqueous solution under visible light (see Table 8).

Table 8. Preparation conditions and removal of pollutants using Nitrogen (N)– or Sulfur (S)– doped TiO$_2$ nanomaterials.

Doping atom	Preparation method	Ti precursor	Dopant source	Removal target	Irradiation wavelength (nm)	Ref.
N	Wet method	TTIP	Ammonia (NH$_4$OH)	CO	433	[403]
N	Heating TiO$_2$	TiO$_2$	Urea and guanidine	2–propanol	420	[404]
N	Sol–gel	Tetra–n–butyl titanium	Ammonium carbonate	Methyl orange and 2–mercaptobenzothiazole	visible–light	[405]
N	Precipitation–hydrothermal	Tetrabutyl titanate	Ammonium chloride	Rhodamine B	≥420	[407]
N	Mixing and calcination at 500°C	TTIP	Thiourea	4–chlorophenol	≥455	[409]
S	Modified sol–gel	TTIP	Ammonium thiocyanate or thiourea	Gaseous acetaldehyde	>420	[419]
S	Mixing and calcination at 400°C	TTIP	Thiourea	Methanol and 4–(methylthio)phenyl methanol	355 and 430	[420]
S	Mixing and calcination at 500–700°C	TTIP	Thiourea	Methylene blue, 2–propanol, adamantane	440	[421]
S	Sol–gel and Grinding anatase with thiourea (calcination at 400°C)	TTIP	Thiourea	Phosphatidylethanolamine lipid	>410	[422]
S	Low–temperature hydrothermal	Titanium disulfide	Titanium disulfide	4–chlorophenol	>400	[423]

Liu *et al.* [405] also prepared yellow nitrogen–doped titania by sol–gel method in mild condition, with the elemental nitrogen source from ammonium carbonate. The analytical results demonstrated that all catalysts were anatase and the crystallite size of nitrogen–doped titania increased with increasing N/Ti ratio. The doping of nitrogen enlarged the specific surface and extended the absorption shoulder into the visible–light region. Photocatalytic activity of the nitrogen–doped titania catalysts was evaluated based on the photodegradation of methyl orange and 2–mercaptobenzothiazole in aqueous solution under visible light (see Table 8). The group stated that the visible–light activity of nitrogen–doped titania was much higher than that of the commercial Degussa P25.

Valentin *et al.* [406] prepared N–doped *TiO₂* samples via the sol–gel method using several nitrogen containing inorganic compounds (e.g. NH_4Cl, NH_3, N_2H_4, NH_4NO_3 and HNO_3) as the nitrogen source. A solution of titanium (IV) isopropoxide in isopropylic alcohol was mixed with an aqueous solution of a nitrogen compound and kept upon constant stirring at room temperature (RT). The solution obtained was left aging overnight at RT to ensure the completion of the hydrolysis. The solution was then dried at 343 K. The dried compound was heated at 773 K in the air for 1 h. They found that the best results were obtained using ammonium chloride as the nitrogen source. The calcination influences the final properties of the material depending on the temperature and heating rate employed in the treatment. After heating in air at 773 K and at a relatively slow heating rate (5 K/min), the final material exhibited a pale yellow color and consisted of anatase structure.

Zhang *et al.* [407] prepared nanoparticles of titanium dioxide co–doped with nitrogen and iron (III) using the homogeneous precipitation–hydrothermal method (see Table 8). They found that the photocatalyst co–doped with nitrogen and 0.5% Fe^{3+} showed the best photocatalytic activity. The degradation efficiencies were improved by 75% and 5% under visible and ultraviolet irradiation, respectively, when compared with the pure titania. It was presumed that the nitrogen and Fe^{3+} ion doping induced the formation of new states closed to the valence band and conduction band, respectively. High photocatalytic activity in the visible light region was also attributed to the cooperation of the nitrogen and Fe^{3+} ion. This cooperation led to a narrowing of the band gap. The co–doping can also promote the separation of the photogenerated electrons and holes to accelerate the transmission of photocurrent carriers.

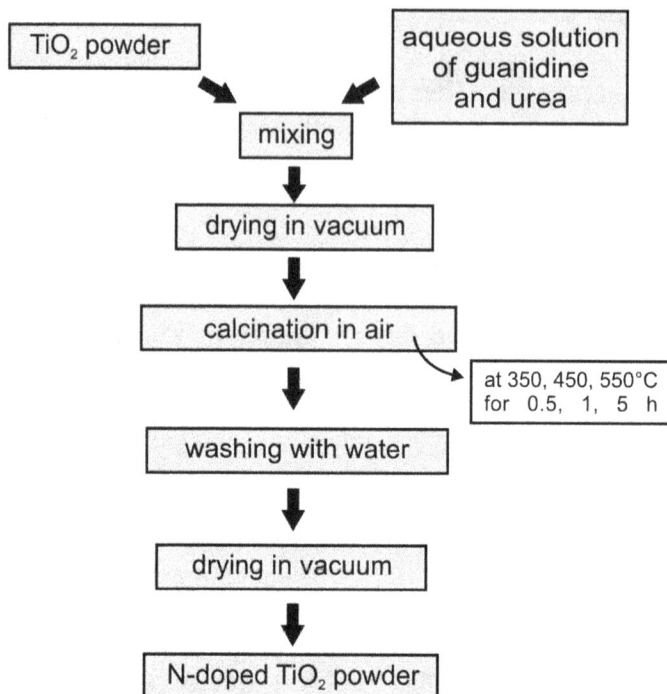

Figure 23. Flow chart for the preparation of nitrogen doped TiO_2 by calcinations with organic nitrogen compounds.

N–doped TiO_2 nanomaterials have also been synthesized by hydrolysis of TTIP ($Ti(OCH(CH_3)_2)_4$) in a water/amine mixture and the post–treatment of the TiO_2 sol with amines [408–412] or directly from a Ti–bipyridine complex [400].

In addition to previously mentioned processes, the production of N–doped TiO_2 nanomaterials has been reported through other methods. These processes are ball milling of TiO_2 in a NH_3 water solution [413], heating TiO_2 under NH_3 flux at 500–600°C [414, 415], calcination of the hydrolysis product of $Ti(SO_4)_2$ with ammonia as precipitator, decomposition of gas–phase $TiCl_4$ with an atmosphere microwave plasma torch [416], ion implantation techniques with nitrogen [417] and N_2^+ gas flux [418] (see Table 8).

S–doped TiO_2 nanomaterials were synthesized by modified sol–gel route [419], mixing titanium tetraisopropoxide (TTIP) with ethanol containing thiourea [420, 421], grinding anatase with thiourea followed by calcination at 400°C [422], heating of mixtures of thiourea and anatase from 200 to 900°C [422], anodization [423], low–temperature hydrothermal method [424], heating sulfide powder [425], or by using sputtering or ion–implanting techniques with S^+ ion flux [426] (see Table 8). Different doping methods can induce the different valence states of the dopants. For example, the incorporated S from thiourea had S^{4+} or S^{6+} state, while direct heating of TiS_2 or sputtering with S^+ induced the S^{2-} anion [419–427].

Ohno *et al.* [421] reported that S–doping shifted the absorption edge of TiO_2 to a lower energy. Thereby exhibits photocatalytic degradation of methylene blue under visible light irradiation. They suggested that sulfur was doped as an anion and replaced the lattice oxygen in TiO_2. On the contrary, reports by Ohno *et al.* [420–422] found that S atoms were incorporated as cations and replaced Ti ions in the sulfur–doped TiO_2.

Jin *et al.* [428] synthesized a novel photocatalyst, carbon–sulfur–codoped TiO_2, by the hydrolysis of tetrabutyl titanate in an aqueous mixture containing thiourea and urea. The co–doped TiO_2 was also prepared by calcining amorphous or anatase TiO_2 with a mixture of thiourea and urea. The photocatalytic activity was evaluated by the photodegradation of 4–chlorophenol under both UV and visible radiation. By investigating the crystal structures, optical properties and photocatalytic activities of various samples, researchers suggest that the wet chemistry process and the crystal transition process from amorphous to anatase are critical in the doping process.

Hamal and Klabunde [419] utilized a modified sol–gel route to synthesize nanoparticle photocatalysts based on silver, carbon and sulfur–doped TiO_2 with a homogeneous anatase crystalline phase and high surface area. The visible light reactivity of the catalyst was evaluated for the photodegradation of gaseous acetaldehyde as a model indoor pollutant. They found that the silver (I) ion, Ag^+, significantly promoted the visible light reactivities of carbon and sulfur–doped TiO_2 catalysts without any phase transformation from anatase to rutile. Moreover, Ag/(C, S)–TiO_2 photocatalysts degraded acetaldehyde 10 times faster in visible light and 3 times faster in UV light than the accredited photocatalyst P25–TiO_2. The visible photoactivities of Ag/(C, S)–TiO_2 were predominantly attributable to an improvement in anatase crystallinity, high surface area, low band gap and nature of precursor materials used.

S–doped titania may be prepared by a method consisting in annealing titanium disulfide at 500°C for 90 min. The resulting powder exhibits in the X–ray photoelectron spectrum (XPS) a sulfur $2p(3/2)$ peak at 160.0 eV and helps in the decolorization of methylene blue by UV and visible light [408]. In another method, titanium tetraisopropoxide was mixed with thiourea in ethanol solution and was followed by evaporation of ethanol under reduced pressure and calcination at 500°C [429]. From the XPS sulfur $2p(3/2)$ peak at 170.0 eV it was concluded that the material contains sulfur in the oxidation state +6. This material photocatalysed methylene blue by visible light ($\lambda \geq 440$ nm) [409, 429].

Boron–doped TiO_2 nanomaterials were synthesized by a sol–gel method using $Ti(OBu)_4$ as titanium source and H_3BO_3 as boron source (see Table 9) [430–432]. 17.0 g of $Ti(OBu)_4$ were dissolved at 25°C in 40.0 mL of anhydrous ethanol under argon atmosphere to form solution 1. 3.0 mL of concentrated HNO_3 were mixed with 35.0 mL of anhydrous ethanol and 15.0 mL of water to prepare the solution 2. Solution 1 was added drop–wise into solution 2 under argon atmosphere and stirred vigorously for 20 min. Appropriate amounts of H_3BO_3 were dissolved in 10.0 mL of bi–distilled water and rapidly added drop–wise to the resulting solution. The solution was continuously stirred for 30–60 min until the formation of TiO_2 gel. After aging for 24 h at room temperature, the as–prepared TiO_2 gel was dried at 120°C for 12 h. The obtained solid was annealed at 450°C for 6 h with a heating rate of 3°C/min [430, 431].

Khan *et al.* synthesized a chemically modified TiO_2 by controlled combustion of Ti metal in a natural gas flame. The flame temperature was maintained close to 850°C. The resulting photocatalyst could absorb UV and most of the visible light below 535 nm [433]. Following Khan's work, Irie *et al.* prepared carbon–doped anatase TiO_2 powders by oxidative annealing titanium carbide (TiC) under O_2 flow at 600°C. Their catalyst showed photocatalytic activities for the decomposition of 2–propanol to CO_2 via acetone under visible light irradiation (400–530 nm) [434]. Sakthivel and Kisch synthesized carbon–doped TiO_2 by the hydrolysis of titanium tetrachloride with tetrabutylammonium hydroxide followed by calcination at 400 and 500°C. In the degradation of 4–chlorophenol by visible light ($\lambda \geq 455$ nm) the catalyst powders have high photocatalytic activities (see Table 9) [435]. Recently, carbon–

doped titania with high surface area ($204 \ m^2/g$) was prepared by temperature–programmed carbonization of anatase under a flow of cyclohexane at temperatures between 450 and 500°C [436]. This carbon–doped titania has much better photocatalytic activity for gas–phase photooxidation of benzene under irradiation of artificial solar light than pure titania (see Table 9). These carbon–doped TiO_2 were all synthesized at high temperature. So it is a challenge to prepare carbon–doped TiO_2 at a low temperature, especially for the energy–saving production of visible–light driven photocatalyst in a large scale for pollutants removal.

Recently, Sun and Li reported an interesting low–temperature method to synthesize carbon nanostructures under mild aqueous condition using glucose as the carbon source [437]. They prepared mesoporous carbon–doped TiO_2 by using glucose and amorphous titanium dioxide. Zhang *et al.* [438] found that the reduction of glucose and crystallization of TiO_2 as well as the carbon doping could take place at the same time under hydrothermal treatment at 160°C. This is one of the lowest temperatures to prepare carbon–doped TiO_2. It was found that the resulting carbon–doped TiO_2 exhibited much higher photocatalytic activity than the undoped counterpart and Degussa P25 on the degradation of Rhodamine B under visible light irradiation (see Table 9).

C–doped TiO_2 was also prepared through the following processes: controlled oxidative annealing of titanium carbide (TiC) for decomposition of trichloroacetic acid under visible light irradiation [439], anodization [423], radio–frequency magnetron sputtering method [440], heating TiO_2 gel in an ionized N_2 gas and then calcination at 500°C [441, 442], modified sol–gel route using titanium(IV) isopropoxide as titanium precursor and ammonium thiocyanate, C₆₀ or thiourea as carbon source [419, 443], hydrolysis of tetrabutyl titanate in a mixed aqueous solution containing thiourea and urea for preparation of visible–light–driven carbon–sulfur–codoped TiO_2 photocatalysts [428], hydrolysis of titanium tetra–*n*–butyl oxide (TTB) in a water/tetrabutylammonium hydroxide (TBAH) mixture and the calcination TiO_2 sol [444].

Table 9. Preparation conditions and removal of pollutants using *NS–TiO₂* doped with Boron (B), Vanadium (V), Carbon (C),

Doping atom	Preparation method	Ti precursor	Dopant source	Removal target	Irradiation wavelength (nm)	Ref.
B	Sol–gel	TBO	H_3BO_3	Methyl orange	360	[430]
B, V	Modified sol–gel	TBO	H_3BO_3 and Vanadium alkoxide	Methylene blue	360	[431]
B	Modified sol–gel	TBO	H_3BO_3	Trichlorophenol, 2,4–dichlorophenol and Sodium benzoate	> 420	[432]
C	Annealing of TiC under O_2	Titanium carbide (TiC)	Titanium carbide (TiC)	2–propanol	400–530	[434]
C	Hydrolysis and calcination at 400 and 500°C	Titanium tetrachloride	Tetrabutylammonium hydroxide	4–chlorophenol	≥ 455	[435]
C	Temperature–programmed carbonization	Anatase titania	Cyclohexane	Gaseous benzene	Solar light	[436]
C	Low Temp. hydrothermal	TTIP	Glucose	Rhodamine B	> 420	[438]
C	Controlled oxidative	Titanium carbide (TiC)	Titanium carbide (TiC)	Trichloroacetic acid	> 410	[439]
F	Mixing TTIP and H₂O–NH₄F	TTIP	NH_4F	Acetone	365	[447]
F	Spray pyrolysis	H_2TiF_6	H_2TiF_6	Gaseous acetaldehyde	> 420	[450]
F	Chemical vapor deposition	Anodized titanium sheet	NaF	Methyl Orange	> 400	[453]
F	Sol–gel–solvothermal	Tetrabutyl titanate	Ammonium fluoride	p–chlorophenol and Rhodamine B	400–500	[455]

F–doped TiO_2 nanomaterials were synthesized through different methods. These processes are mixing tetraisopropoxide with ethanol containing H_2O–NH_4F, [445– 447], heating TiO_2 under hydrogen fluoride [448, 449], spray pyrolysis from an aqueous solution of H_2TiF_6 [450, 451], using ion–implanting techniques with F^+ ion flux [452], chemical vapor deposition of anodized Ti in $C_2H_2O_4 \cdot 2H_2O + NH_4F$ electrolyte [453], sol–gel method using tetraethyl orthotitanate as titanium precursor and CF_3COOH as fluorine source [454], sol–gel–solvothermal method using tetrabutyl titanate and ammonium fluoride as precursors [455] (see Table 9).

Chapter 5

Supported and Immobilized Titanium Dioxide Nanomaterials

The most widely used photocatalytic process, in the literature, is carried out in a discontinuous slurry photoreactor operating with titanium dioxide suspensions. However, slurry reactors have a number of practical and economical disadvantages. The main problem related to suspended photocatalyst systems is the separation of $NS–TiO_2$ after treatment. As TiO_2 materials are usually non–porous, to maximize their activity, particles should be small enough to offer a high specific surface area, which imposes high filtration costs. Moreover, the recent studies have raised concerns about the potential toxicity of titanium dioxide nanoparticles [456]. Supported photocatalysts have been developed in an attempt to solve this problem. The most important properties of a suitable support are its being chemically inert, presenting a high specific surface area and its transparency to UV radiation.

The main advantage of immobilized–TiO_2 photocatalytic reactors is their application in continuous treatment of contaminated water. Immobilization procedure of $NS–TiO_2$ must guarantee the long-term stability and avoid possible leaching of TiO_2 particles to the solution. It must also allow regeneration of $NS–TiO_2$ in case of deactivation. This point assumes even greater significance in the case of real wastewater treatment. The chemical composition of wastewater has a strong influence on photocatalytic efficiency. In this chapter, we are to describe the immobilization methods of $NS–TiO_2$ on different substrates. Different methods for immobilization of $NS–TiO_2$ on solid support substrates have been listed in Tables 10–12 & 14–20.

5.1. Immobilization on Glass Substrates

There are extensive references about the immobilization of $NS–TiO_2$ on the different kinds and forms of glass substrates in the literature [465–460]. One of the advantages of the glass supports in comparison with other supports, such as polymeric materials is that the glass substrates are inert and non–degradable under the photocatalytic process. The immobilization methods of $NS–TiO_2$ on the different kinds of glass substrates are presented in Table 10.

The heat attachment method is made use of to immobilize $NS–TiO_2$ on glass beads according to the litrurure [41, 56, 458]. In this method, glass beads are etched with dilute hydrofluoric acid (5% v/v) for 24 h and washed thoroughly with deionized water. It makes a rough surface for a better contact surface between $NS–TiO_2$ and the glass surface. A TiO_2 slurry is prepared through sonication of a mixture containing 1.5 g TiO_2 in 200 mL of deionized water. The glass beads, immersed in the slurry of TiO_2, are thoroughly mixed for 20 min and then removed from the suspension and eventually placed in an oven for 1.5 h at 150°C. They are, subsequently, placed in a furnace for 2 h at 500°C. The samples are thoroughly washed with double distilled water to remove free TiO_2 particles (see Figure 24). Photocatalytic removal of textile dyes, Basic Red 46 and Direct Red 23, has been tested using these glass beads (see Table 10) [41, 56].

In addition, the heat attachment method was used to fix $NS–TiO_2$ on glass plates [458]. The process is carried out in certain stages; a suspension of Millennium PC–500 TiO_2 of 4 g/L in deionized water is prepared. The suspension concentration is chosen so as to get thin enough deposits. The pH is normally adjusted to about 3 using diluted HNO_3 and the suspension is sonicated for 15 min. Then proper volume of suspension is carefully poured on the glass plates and allowed to dry out at room temperature for 12 h. Then the plates are completely dried out at 100°C for an hour. Having been dried, the plates are calcined at 475°C for 4 h. Before deposition, the glass surface is washed in a basic solution of NaOH in order to increase the number of OH groups. As shown in Figure 25, the first coat is not capable of covering the entire surface, a complete coverage is accomplished by additional coats. Therefore, this deposition process is carried out three times in a row so as to increase the total thickness (see Figure 25). The plates are thoroughly washed with deionized water for removal of the free TiO_2 particles. Photocatalytic degradation of three commercial textile dyes (i.e. Acid Orange 10, Acid Orange 12 and Acid Orange 8) using immobilized TiO_2 nanoparticles on glass plates in a circulation photoreactor has been investigated [458].

Having explained the heat attachment method, let's proceed to investigate dip–coating method for immobilization of $NS-TiO_2$ on hollow Pyrex glass beads with average diameter of 20 mm. After being carefully cleaned by sonication in acetone, the beads are immersed in a solution of 0.1 M titanium tetraisopropoxide in ethanol (200 mL) and hydrochloric acid (2 N, 5.4 mL). The hollow Pyrex glass beads are removed from the solution at a constant rate of 2 cm/min. The samples are dried in the air for 15 min and calcined at 400°C for 2 h. Photocatalytic inactivation of three species of algae (i.e. *Anabaena, Microcystis* and *Melosira*) has been carried out with the TiO_2–coated Pyrex hollow glass beads under the illumination of UV–A light. After being irradiated with UV–A light in the presence of the TiO_2–coated Pyrex glass beads, *Anabaena* and *Microcystis* loss their photosynthetic activity. The string of *Anabaena* cells and the colonies of *Microcystis* cells are completely separated into individual spherical one [457].

TiO_2 nanoparticles are supported on glass Raschig rings (8 mm long × 7 mm o.d.) by the repeated dip–coating method, air drying and calcination at 400°C for 10 min several times. The catalyst Raschig rings are thoroughly washed with deionized water under stirring, so that the possibility of TiO_2 particles leaching to the irradiated solution during the reaction is avoided. Photodegradation of naphthalene in water using TiO_2 supported on glass Raschig rings in continuously stirred tank reactor has been reported [459].

Nano–composites of Fe–doped TiO_2, immobilized on aluminosilicate hollow glass microbeads (HGMBs), are prepared by co–thermal hydrolysis deposition of titanium sulfate and iron nitrate in hot acidic water, followed by calcinination. In a typical experiment, firstly, 0.5 g HGMBs and 1 mL (0.1 M) sodium dodecyl sulfate organic ligand are added into a 100 mL beaker containing 20 mL H_2O as the stabilizing agent. The pH of the solution is then modulated to 2.0–3.0 by H_2SO_4. Secondly, 6 mmol $Ti(SO_4)_2$ and an appropriate amount of $Fe(NO_3)_3$ are dissolved in 20 mL H_2O, which are dripped to the above beaker at 100°C. After the reaction, the composite product is filtrated, washed with deionized water and dried at 120°C for 2 h, to be calcined at 500°C for another 2 h. Photocatalytic activity of Fe–doped TiO_2 nano–composites, deposited on HGMBs, has been tested in photodegradation of methyl orange under visible light irradiation [461].

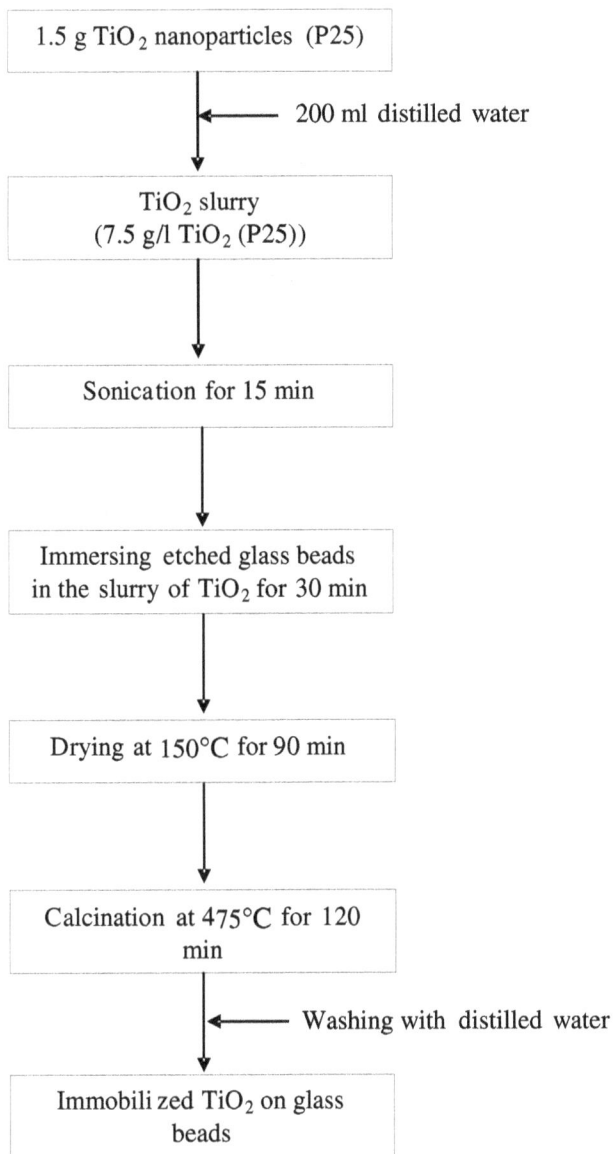

Figure 24. The procedure of immobilization of TiO_2 nanoparticles on glass beads.

Figure 25. Scanning electron microscopy images of *TiO₂* nanoparticles deposited on glass plates: a) First coat, b) Second coat, c) Third coat. (Adapted from Khataee *et al.* [458] with permission from publisher, Elsevier. License Number: 2627050906208).

Miki–Yoshida *et al.* reported the preparation of NS–TiO_2 thin films inside borosilicate glass tubes by spray pyrolysis technique [462]. The overall dimensions of the tubes included an internal diameter of 22 mm and a length of 120 cm. The borosilicate glass tube had been coupled to a medical nebulizer, which was used as an atomizer. A three–zone cylindrical furnace heated this tube, with a precise temperature control (± 1 K). The starting solution was a 0.1 mol/dm^3 of titanyl acetylacetonate in absolute ethanol. The process started with the aerosol generation of precursor solution in the nebulizer. This aerosol was subsequently conveyed by the carrier gas and injected directly into the heated tube, inside the cylindrical furnace. The carrier gas was micro–filtered air, the pressure was kept at 310 kPa and the flux was controlled with a mass flow control between 142 and 250 cm^3/s. All the samples were prepared through intermittent spraying to improve film–thickness uniformity and the overall quality. During the rest period, a ventilation flow was maintained. The spraying time varied between 60 and 120 s and the rest time was 300 s. After deposition, all the films were heated in the air at 452°C for 2 h so that all organic residues deposited in the surface would decompose and the films' microstructure would stabilize. Finally, the samples were left to cool down inside the furnace at the room temperature.

Indium–tin oxide (ITO) glasses have also been used as support for NS–TiO_2 [464–467]. For instance, Peralta–Hernández *et al.* [464] reported deposition of TiO_2–carbon nanocomposite on ITO glass plates by electrophoretic deposition (ED) method. ITO glass plates were immersed in 10 mL of a colloidal suspension of TiO_2–carbon nanocomposite particles. Accordingly, a 4 V potential difference was applied between a stainless steel shield and the negative ITO plate for a period of 40 s at room temperature. The distance between the electrodes was 2 cm. Fresh electrodes were placed in an oven to sinter the nanocomposite film in the air at 450°C for 30 min. Photocatalytic activity of prepared electrodes were tested for removal of Orange II [464].

Mansilla *et al.* deposited NS–TiO_2 on sintered glass cylinders [470]. In order to prepare the support material, Pyrex glass was mechanically crushed, sieved to obtain particles of 150–600 μm and then were poured into a cylindrical ring ceramic mould. The mould was formed with two concentric cylinders of refractory and highly temperature–resistant ceramic. The heating temperature (>700°C) was chosen to allow the glass particles to be fused, thereby avoiding material melting. The size of each fritted cylinder obtained after 2 h in the oven was 5.8 cm length, 3.1 cm and 4.1 cm of internal and external diameters, respectively. The thickness of each cylinder wall was 5 mm. The impregnation of *NS–TiO$_2$* (Degussa P25) on sintered glass cylinders was carried out, with each cylinder submerged in 60 mL of slurry *TiO$_2$* during 20 min. The heterogeneous titania solution was prepared by the mixture of 42 mL of distilled water, 18 mL of ethanol and 3 g of Degussa P25. The cylinders were first dried at room temperature and then heated using a temperature program during 4 h at 280°C. Finally, the impregnated cylinders were heated for 3 h at 400°C. The *TiO$_2$* –coated cylinders were sonicated for 30 min before their use in catalytic reactions. The antibiotic flumequine was used to evaluate the photocatalytic activity of *NS–TiO$_2$* coating on sintered glass cylinders in an annular photoreactor.

Glass fiber was reported as an appropriate support for immobilization of *TiO$_2$* nanoparticles [472–475]. Scotti *et al.* explained the immobilization of *TiO$_2$* nanoparticles on glass fiber for photocatalytic degradation of phenol (see Table 10) [472]. 1.0 g *TiO$_2$* powder was suspended by sonication in 13.2 mL of water. Polyethylene aqueous solution (5.0 g, 50 wt%) and a few drops of Triton X–100, while stirring, had sequentially added. So that, a suitable viscosity for a good and uniform coating adhesion could be obtained. The mixture was homogeneously pasted on the glass fiber material (size 7 cm × 20 cm) with a brush and left to dry in an oven at 120°C for 30 min. The coating was calcined at 400°C in the air for 30 min for the organic content to be fully removed.

Table 10. Immobilization of $NS–TiO_2$ on glass substrates for photocatalytic treatment of contaminated water.

Kind and shape of glass support	Immobilization method	Removal target	Ref.
Glass beads	Heat attachment	Basic Red 46 and Direct Red 23	[41, 56]
Hollow glass beads	Dip–coating	Three species of algae: *Anabaena, Microcystis and Melosira*	[457]
Glass plate	Heat attachment	Three orange dye: Acid Orange 10, Acid Orange 12 and Acid Orange 8	[458]
Glass Raschig rings	Dip–coating, air drying and calcination at 400°C for 10 min	Naphthalene	[459]
Aluminosilicate hollow glass microbeads	Co–thermal hydrolysis deposition in the acidic water solution and calcination at 500°C for 2 h	Methyl Orange	[461]
Borosilicate glass	Dip–coating	Glucose	[463]
Indium–tin oxide (ITO) glass	Electrophoretic deposition	Orange (II)	[464]
Quartz	Dip–coating	Malic acid	[468]
Glass drum	Double spread method	Phenol	[469]
Sintered glass cylinders	Heat attachment	Antibiotic: Flumequine	[470]
Glass tubes	Dip–coating	Pesticide: Paraquat	[471]
Glass fibers	Polyethylene glycol and Triton X–100 as coating adhesion; and sol–gel method	Phenol and Benzamide	[472, 476, 477]

5.2. Immobilization on Stone, Ceramic, Cement and Zeolite

Inorganic substrates such as Perlite, Pumic stone, porous Lava, zeolites, cements pellets and ceramic tiles have a high specific surface area and photocatalytic resistance. These materials have been used as supports for immobilization of $NS-TiO_2$ (see Tables 11, 12, 14) [478–482]. In this section, we are to describe the immobilization of $NS-TiO_2$ on the different inorganic supports.

Rego *et al.* [478] deposited TiO_2 and ZnO particles on common ceramic tiles, by screen–printing technique. In order to deposit ZnO and TiO_2, the powders were suspended (1:1 wt.%) in an organic medium (NF 1281) and the layers were printed through distinct sieved screens (136 mm) on common, bright monoporosa glaze tiles. The deposited layers were then fired at 850°C. These layers were characterized and tested for the photocatalytic degradation of Orange II in aqueous solutions under sunlight [483, 484].

Plesch *et al.* [485] prepared ceramic macroporous reticular alumina foams with a pore size of 15, 20 and 25 pores per inch. $NS-TiO_2$ thick films were supported on the foam surface by a wash–coating process. For this purpose, 20 wt.% of TiO_2 powder was stirred in distilled water and acidified with 10% HNO_3 to obtain a pH value of 2.4. The viscosity was adjusted by the addition of small amounts of water. The TiO_2 coating was performed by dipping the pre–sintered alumina foams into the titania slurry for 5 min, followed by a 10 min ultrasonic bath treatment to give a better homogeneity of the slurry. $NS-TiO_2$ coated samples were dried at 80°C for 10 min, followed by a heat treatment at 600°C for 1 h with a heating and cooling rate of 60°C/h. Photocatalytic degradation of phenol was studied in the presence of $NS-TiO_2$ coated ceramic.

Zeolites seem to be promising supports for $NS-TiO_2$ because of their regular pores and channel sizes, high surface area, hydrophobic and hydrophilic properties, easily tunable chemical properties, high thermal stability, eco–friendly nature and good adsorption ability. Photocatalytic activity of supported TiO_2 on zeolite enhances by high adsorption capacity of zeolite. It has been indicated that natural zeolites, as well as synthetic zeolites such as HZSM–5, ZSM–5, 13X, 4A, β, HY, Hβ, USY,

Y zeolites, are effective supports for TiO_2 photocatalyst (see Table 11) [486–499].

Highly dispersed titanium oxides, included within zeolite cavities, are prepared using an ion–exchange method and used as the photocatalyst for the direct decomposition of NO at 2°C [500, 501]. These catalysts, having a tetrahedral coordination, show a high photocatalytic activity compared with that of titanium oxide catalysts prepared by an impregnation method, as well as with that of a bulk TiO_2 powder catalyst. It was indicated that a high photocatalytic efficiency and selectivity for the formation of N_2 in the photocatalytic decomposition of NO was achieved with TiO_2/Y–zeolite catalyst having a highly dispersed isolated tetrahedral titanium oxide species [501].

Reddy *et al.* [493] reported photocatalytic degradation of salicylic acid using TiO_2/Hβ zeolite. TiO_2 is well dispersed over Hβ zeolite at moderate loading which prevents particles from aggregating and the light from scattering. The fine dispersion of TiO_2 on Hβ zeolite leaves more number of active sites near the adsorbed pollutant molecules, which results in fast degradation. Further, the strong electric field present in the zeolitic framework can effectively separate the electrons and holes produced during photoexcitation of TiO_2. Sankararaman *et al.* reported the ability of zeolites favoring photo–induced electron transfer reactions and retarding undesired back electron transfer [502]. They found that Hβ zeolite increased the adsorption of pollutants and generated a large amounts of hydroxyl and peroxide radicals, which are critical species in the photocatalytic degradation process. Zeolites can delocalize excited electrons of TiO_2 and minimize e^-/h^+ recombination. TiO_2/zeolite shows enhanced photodegradation due to its high adsorption property by which the pollutant molecules are pooled closely and degraded effectively [503].

TiO_2/Hβ with different wt.% of TiO_2 are prepared through adding an appropriate amount of TiO_2 and 1 g of Hβ in acetone [504]. The mixture is magnetically stirred for 8 h at ambient temperature. The mixture is then filtered, dried at 110°C for 3 h and calcined in the air at 550°C for 6 h. The prepared zeolite–supported TiO_2 has been used for photocatalytic degradation of a pesticide propoxur (see Table 11).

Table 11. Photocatalytic removal of pollutants using supported NS–TiO_2 on zeolites.

Type of zeolite	Type and crystal size (nm) of TiO_2	TiO_2 loading (wt.%)	Removal target	Initial concentration	Catalyst dosage	Contact time (min)	Removal efficiency (%)	Ref.
ZSM–5 zeolite	Anatase (Merck)	2	EDTA	5 mm	4 g/L	60	99.9	[488]
HZSM–5 zeolite	Anatase, 27	15	Phenol and p–chlorophenol	1 and 0.1 mm	75 mg/25 mL	30	>95	[490]
USY and β-zeolites	Hombikat (UV–100), <10	25	Salicylic acid	2 mm	–	–	–	[493]
Natural zeolite, Clinoptilolite ($Na_8[Al_8Si_{40}O_{96}]\cdot32H_2O$)	Synthesized, 80	5	Methyl orange	16 mg/L	40 mg/10 mL	94	95	[498]
Hβ zeolite	Degussa P25, 21	20	Propoxur	200 mg/L	100 mg/100 mL	360	>95	[504]
Y zeolite	Anatase, 8–30	20	Basic violet 10	10 mg/L	5333 mg/L	240	>95	[505]
HY zeolite	Degussa P25, 25	1	2,4–dichlorophenoxy acetic acid (2,4–D)	200 mg/L	200 mg/100 mL	540	100	[506]
NaA zeolite	Degussa P25, 25	10	Methylene blue	10 mg/L	0.1 g/150 mL	60	>95	[507]
Natural zeolite, Mordenite ($Na_8[Al_8Si_{40}O_{96}]\cdot24H_2O$)	Synthesized, 80	5	Methyl orange	16 mg/L	40 mg/10 mL	90	100	[508]

Table 12. Immobilization of $NS–TiO_2$ on ceramic, stone and brick for photocatalytic applications.

Support substrate	Immobilization method	Photocatalytic application	Ref.
Ceramic tiles	Screen–printing	Removal of Orange II	[478, 484]
Ceramic ($ZrO_2 + ZrSiO_4$)	Sol–gel dip–coating	Removal of Methyl Orange	[479]
Pumice stone	Pasted by cement or polycarbonate	Acid orange 7 and 3–nitrobenzenesulfonic acid	[509]
Pumice stone	Brushing with TiO_2 milk or impregnating TiO_2 milk with conventional soaking, drying and heat treatment methods	Photocatalytic disinfection and detoxification of *E. coli*, Acid Orange 7, Resorcinol, 4, 6–dinitro–*o*–cresol, 4–nitrotoluene–2–sulfonicacid, Isoproturan	[510]
Porous lava	20–100 g/L TiO_2 slurry is impregnated on a slice of 14 cm^2 Volvic lava with a brush. The coated support is then subjected to reduced pressure (100 mbar) for 1 min. Then, it is dried overnight at 100°C	Removal of 3–nitrobenzenesulfonic acid	[511]
Red brick	Sol–gel dip–coating	Removal of 3–nitrobenzenesulfonic acid	[511]
Black sand	Titanium tetra–isopropoxide (1.5 mL) is dissolved in isopropanol (40.0 mL). Pre–dried Si/Black sand particles (0.2 g) are dispersed in the solution and sonicated for 0.5 h. Then, it is stirred for 5. The mixture is heated at 80°C for 3 h and then calcined at 450°C for 3 h	Removal of six dyes: Direct red 80, Eosin B, Rose bengal, Orange II, Ethyl violet, Rhodamine B	[512]
Sand from a coastal dune	Preparing a suspension of 0.5 g TiO_2 in 100 mL water, adding 100 g of sand to the sonicated suspension and evaporating to dryness in an oven at 100°C. The coated sand was then heated at 550°C for 0.5 h	Removal of Methylene blue, Rhodamine B, Methyl orange and salicylic acid	[513]

Li *et al.* [498] reported preparation of TiO_2/zeolite from precursors of $Ti(OC_4H_9)_4$ and natural zeolite, clinoptilolite. The researchers indicated that the synthesized TiO_2/zeolite displays higher photocatalytic activity than pure TiO_2 nanopowders in degradation of methyl orange. The reason of this observation was concluded to be due to the super adsorption capability of the zeolite support.

In recent years, the natural mineral materials such as pumice stone and Volvic Lava, have also been used as supports of $NS–TiO_2$ due to their layered or porous structure, low cost and abundant storage [509–511]. Photocatalytic disinfection of real river waters containing *E. coli* with $NS–TiO_2$ immobilized on pumice stone was studied. The supported $NS–TiO_2$ on pumice stone was also used to remove different organic substrate like acid orange 7, resorcinol, 4, 6–dinitro–o–cresol, 4–nitrotoluene–2–sulfonicacid, isoproturan [510] (see Table 12).

Recently, magnetic supports have been proposed for immobilization of $NS–TiO_2$ [512–515]. Magnetic supports are a type of composite often composed of a TiO_2 shell, an insulating silica layer and a magnetic core that makes the photocatalyst recoverable using an external magnetic field. Natural magnetic black sand was used as the core for the preparation of a magnetic photocatalyst. The average size of the black sand was 10 μm, which can render a significant interaction with the magnetic field. A rough silica layer and titanium dioxide layer were then deposited on the black sand to form a magnetic photocatalyst (see Figure 26). The prepared magnetic photocatalyst was used to remove a series of laboratory dyestuffs [512] (see Table 12).

Furthermore, perlite has been reported to be a suitable support for titanium dioxide nanomaterials [516–520]. Basically an amorphous alumina silicate (see Table 13), Perlite is an industrial mineral and a commercial product, useful for its light weight after processing. Due to its low density and relatively low price, many commercial applications for perlite have been developed including construction and manufacturing fields, horticultural aggregates, filter aid and fillers [519].

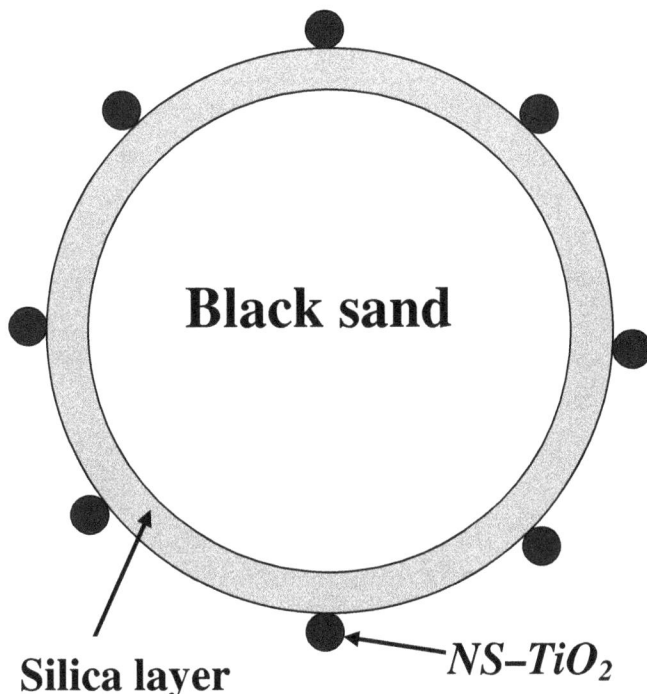

Figure 26. *TiO₂* nanomaterials immobilized on black sand covered by a silica layer.

In order to immobilize *NS–TiO₂* on perlite, 0.5 g of Degussa P25 powder is added to 18 mL ethanol. Then, 1.5 mL of dilute nitric acid with a pH of 3.5 is added to slurry in order to disperse *TiO₂* nanoparticles adequately. Then, the slurry is sonicated for 5 min. 1 g of perlite granules is added to the slurry. Perlite granules are mixed in *TiO₂* slurry for 30 min. Perlite granules which have adsorbed enough *TiO₂* are filtered from the slurry and calcined at 450°C for 30 min. The photocatalytic activity of prepared catalyst has been tested through the degradation of phenol and furfural from aqueous solution [516, 517] (see Table 14).

The cements containing active photocatalytic titania nanoparticles have widespread applications to create environmentally clean surfaces. These applications include self–cleaning surfaces, anti–soiling, de–pollution of VOCs and NO_X contaminants and antifungal/microbial activities [521–528]. The relevant photocatalytic processes may occur both at the air–solid interface and at the liquid–solid interface.

Table 13. Chemical composition of perlite [519].

Constituent	Percentage present
SiO_2	71–75
Al_2O_3	12.5–18
Na_2O	2.9–4.0
K_2O	4.0–5.0
CaO	0.5–2.0
Fe_2O_3	0.1–1.5
MgO	0.03–0.5
TiO_2	0.03–0.2
MnO_2	0.0–0.1
SO_3	0.0–0.1
FeO	0.0–0.1
Ba	0.0–0.1
PbO	0.0–0.5
Cr	0.0–0.1

Lackhoff *et al.* reported the modification of Portland cement with commercially available *NS–TiO₂* samples. The used *NS–TiO₂* samples were Degussa P25 (70% Anatase, 30% rutile; BET 50 m²/g; primary crystal size 21 nm), *TiO₂* Hombikat UV–100 (Anatase; BET>250 m²/g; primary crystal size <10 nm) and *TiO₂* anatase Jenapharm (Anatase; primary crystal size 500 nm). 10 or 20 g of white cement and 5 or 10 wt.% of *NS–TiO₂* were weighed into small plastic containers and mixed as dry powders to produce modified cement. After that, the volumetrically determined amount of water was added (water/cement ratio on all pastes: 0.5) and the samples were again thoroughly mixed. The resulting pastes were filled into tightly closed cylindrical plastic containers made from modified disposable syringes in which they were allowed to hydrate for at least 28 days. After that, the samples were opened and the bulk cement was crushed with a pestle and mortar. The size distribution range of modified cement particles was 1 to 400 μm with the mean values raging from 4 to 8 μm. The photocatalytic activity of the modified cement particles was tested for removal of atrazine under sun light [522] (see Table 14).

Table 14. Immobilization of $NS-TiO_2$ on perlite and cement for photocatalytic applications.

Support substrate	Immobilization and preparation methods	Photocatalytic application	Ref.
Perlite	0.5 g of Degussa P25 is added to 18 mL ethanol, acidified and sonicated. Perlite granules mixed in TiO_2 slurry for 30 min. Then, calcined at 450°C for 30 min.	Degradation of phenol and furfural	[516, 517]
White cement	Sol–gel dip-coating	Removal of 3–nitrobenzenesulfonic acid	[511]
Portland cement	See the text	Removal of Atrazine from water	[522]
Portland cement	Mixing TiO_2 with Portland cement in the ratio 1:1 or 1:2 by grinding. Adding distilled water. Pasting semisolid mass to the inner surface of round tile plates (12 cm diameter) or Petri dish (15.5 × 1.8 cm). Dyeing for 24 h	Degradation of Alphazurine FG and reactive yellow 17 dyes in aqueous solution	[523, 525]
White cement	Samples of white cement having a thickness of 2 mm were placed on supports having a discoid form with a diameter of 3.2 cm and a thickness of 7 mm, containing 5% of TiO_2 (Degussa P25)	Removal of NO_X and BTXE (Benzene, Toluene, Ethylbenzene and Xylene)	[526, 528]
Portland cement	Mixing 0, 1, 3 and 5 wt% of TiO_2 with cement. $NS-TiO_2$ was PC–105 from Millennium (anatase, BET 5±10 m²/g, a crystallite size of 15–25 nm). The cement pastes were mixed with a water/cement ratio of 0.33. After mixing, the fresh materials were cast into 20 × 20 ×1 cm moulds made of expanded polystyrene	Decolorization of Rhodamine B	[529]
Portland cement	Enriching the following substrates with TiO_2 is performed using dip–coating and sol–gel methods.: Concrete and plaster materials mainly used for wall and floor covering; commercial autoclaved white concrete material marketed for nonstructural partition wall and concrete tiles manufactured with different finishing techniques.	Removal of toluene form air	[527]

5.3. Immobilization on Metallic and Metal Oxide Materials

Immobilization of $NS-TiO_2$ on metallic materials such as stainless steel and aluminium plates opens up many new potential uses, because of their mechanical robustness, corrosion resistance, strength and shape–forming properties. These include architectural self–cleaning applications, anti–microbial coatings and many environmental applications including water and air purification [4, 5, 531] (see Tables 15–17). In this section, we describe the immobilization of $NS-TiO_2$ on metallic materials as well as metal oxide substrates.

The main deposition methods of $NS-TiO_2$ onto stainless steels are cathodic arc deposition [532], electrophoretic deposition [533], anodic spark deposition (ASD) [534], the sol–gel method [535, 536], atmospheric pressure metal organic CVD (AP–MOCVD) [537, 538] and radio frequency (RF) magnetron sputtering [539]. Preparation methods and applications of $NS-TiO_2$ thin films on stainless steel have been summarized in Table 15.

Table 15. Preparation of $NS-TiO_2$ thin films on stainless steel and its application.

Preparation method	Application	Ref.
Cathodic arc deposition	Bactericidal effect: removal of *Escherichia coli K12*	[532]
Anodic spark deposition	Photocatalytic degradation of aqueous solution of 4–chlorophenol	[534]
Sol–gel	Photocatalytic degradation of aqueous solution of 4–chlorobenzoic acid	[535, 536]
Atmospheric pressure metal organic CVD	Photocatalytic degradation of stearic acid	[538]
Radio frequency magnetron sputtering	Decomposition of methylene blue in an aqueous solution	[539]
Sol–gel and dip–coating	Corrosion protection	[540]
Spray–drying	Photocatalysis of phenol and methylene blue	[541]
Electrophoretic deposition	Dye–sensitized solar cells	[542]

Lim *et al.* reported application of immobilized TiO_2 nanofibers on titanium plates for modification of dental and bone implants to improve osseo–integration [543]. Titanium and its alloys have been widely used as orthopedic and dental implant materials because of their compatible mechanical properties and good biocompatibility [544, 545]. The researchers fabricated TiO_2 nanofibers by an electrospinning method using a mixture of Ti(IV)isopropoxide and polyvinyl pyrrolidone (PVP) in an acidic alcohol solution. The synthesized nanofibers were immobilized on NaOH/HCl–treated titanium plates by inducing alcohol–condensation reaction of Ti(IV)isopropoxide with Ti–OH group on titanium surface and its subsequent calcination (500–1000°C). The diameter of TiO_2 nanofibers could range from 20 to 350 nm. TiO_2 nanofibers on titanium plates were used for the surface modification of titanium implants to improve the osseo–integration.

Furthermore, TiO_2 powder can be directly immobilized on aluminium plate by electrophoretic deposition technique (see Table 17). The Al plate is properly cleaned with diluted acid (0.1 M H_2SO_4) before being placed in a suspension of TiO_2 in acetone (1 g in 100 mL). The suspension is first homogenized by sonication for 15 min. A counter-electrode of stainless steel is placed in the suspension just in front of the Al plate (as the cathode) with the same size and shape. A potential of 110 V is applied between the electrodes to obtain the deposit. Due to the natural surface charge, TiO_2 particles move to the conducting substrate, thus forming a layer. The substrate is then removed from the suspension and is let to dry at room temperature before weighing it. Coating–drying–weighing process is repeated until a desired amount of TiO_2 is actually obtained on the surface of the substrate. The immobilized TiO_2 has been tested for degradation of phenol [546].

Silica–supported TiO_2 materials have been used to improve the recovery properties of NS–TiO_2 while an acceptable level of photoactivity is maintained [547–552]. For example, Lopez–Munoz *et al.* used two different powdered silica materials as supports for the preparation of TiO_2/SiO_2 photocatalysts [547]. The first one was an amorphous commercial SiO_2 (Grace Sylopol 2104). The second one was a mesostructured silica material called SBA–15 with a very well defined pore size around 6 nm. The latter silica was synthesized using Pluronic 123 triblock copolymer (EO_{20}–PO_{70}–EO_{20}) as template to produce the well–ordered hexagonal mesoporous structure of this material [548]. The surfactant was dissolved under stirring in HCl 1.9 M (weight ratio HCl:Pluronic of 31.25) at room temperature. The solution was then heated up to 40°C before the addition of tetraethylorthosilicate (TEOS, weight ratio Pluronic:TEOS of 2.18). The resultant solution was vigorously stirred for 20 h at 40°C, followed by ageing in a closed recipient at 100°C for 24 h under static conditions. The solid product was recovered by filtration, dried at room temperature and then calcined at 550°C [547, 548]. Silica–supported TiO_2/SiO_2 materials were obtained by adding a solution with the required amount of titanium tetraisopropoxide (TTIP) in isopropanol (TTIP:i–PrOH weight ratio 1:8) to the silica suspension in isopropanol. The suspension was stirred for 45 min at room temperature to allow the diffusion of TTIP inside the porous structure of the silica particles. Then, water was added with a H_2O:TTIP molar ratio of 160 to produce a rapid condensation of TiO_2 inside the pores, while the stirring was being maintained for another 45 min. The solids were recovered from the mixture through centrifuging. Then, it was charged into a Teflon–line autoclave reactor at 110°C for 24 h. It was dried at 110°C overnight and calcined at 550°C for 5 h. The synthesis procedure is schematized in Figure 27.

The photocatalytic activity of silica–supported TiO_2 materials has been evaluated with different model pollutants, such as potassium cyanide, metal–complexed cyanides, alcohols, organic dyes, pesticides and organochloride compounds (see Table 16). The photocatalytic experiments have been carried out in different photocatalytic reactors, using both UV lamps and the solar light [549–552].

Wang *et al.* reported preparation of silica gel–supported TiO_2 by post–synthesis hydrolytic restructuring method [553]. A silica gel with the particle size ranging from 76 to 154 μm and surface area of 415 m^2/g was used. Samples of TiO_2/silica gel were prepared using $Ti(OC_4H_9)_4$ and CH_3CH_2OH as solvent through three steps. Firstly, 10 g silica gel was washed by distilled water and dried overnight. It was, then, transferred into a beaker, containing 50 mL CH_3CH_2OH. The slurry was stirred with known amounts of $Ti(OC_4H_9)_4$. The hydrolysis and condensation of $Ti(OC_4H_9)_4$ and –Si–OH occurred at the surface of silica gel (see Figure 28). This step should be performed in absolute ethanol to avoid the hydrolysis of TiO_2. Secondly, the silica gel containing titanium was separated and washed by CH_3CH_2OH several times. Subsequently, the sample was washed by distilled water and $-Ti(OC_4H_9)_2$ was hydrolyzed to $Ti(OH)_2$. Thirdly, the sample was dried overnight at 120°C and calcined in an oven at 520°C for 30 min to obtain single–layer TiO_2–modified silica gel. If the above experiments were to be repeated, the sample of double–layered TiO_2–modified silica gel could be obtained. Photocatalytic degradation of 20 mg/L methylene blue in the presence of 3 g/L silica gel–supported TiO_2 was 90.96% at the reaction time of 60 min.

Thereafter, it can be concluded that the large surface area and the preferential adsorption of the reacted molecules on SiO_2 promote the photocatalytic activity.

Transitional aluminas, mainly γ–Al_2O_3, are known to have a high surface area, strong Lewis acid centers and remarkable mechanical resistance. Aluminum oxide is widely used as a support of catalysts such as NS–TiO_2 (see Table 17).

Chen *et al.* [563] reported immobilization of TiO_2 on γ–Al_2O_3 for photocatalytic decolorization of methyl orange in aqueous medium. γ–Al_2O_3 has a surface area and a pore volume of 162 m^2/g and 0.45 cm^3/g, respectively. A cursory glance at his work showed that acetylacetone was added to a 0.5 M solution of titanium isopropoxide in ethanol. The molar ratio of acetylacetone to titanium isopropoxide was 1. The solution was stirred at room temperature for 3 h. H_2O was dripped to the solution under vigorous stirring. This sol was stirred at room temperature for 3 h, after which γ–Al_2O_3 powder was added slowly with vigorous stirring. The final solution was kept on stirring at 50°C for 24 h. The solid samples were filtered, washed by water, dried in the air at 100°C for 24 h and calcined at 600°C for 3 h.

```
┌──────────────────────────────────────┐
│  Titanium tetraisopropoxide in       │
│  isopropanol (ratio 1:8)             │
└──────────────────────────────────────┘
                   │
                   ▼
┌──────────────────────────────────────┐
│  SiO₂ in isopropanol (ratio 1:8)     │
└──────────────────────────────────────┘
                   │
                   ▼
┌──────────────────────────────────────┐
│  Stirred for 45 min at 25°C          │
└──────────────────────────────────────┘
                   │           Water
                   ▼
┌──────────────────────────────────────┐
│  H₂O:TTIP molar ratio: 160,          │
│  Stirred for 45 min at 25°C          │
└──────────────────────────────────────┘
                   │           Centrifuging
                   ▼
┌──────────────────────────────────────┐
│  Autoclave at 110°C for 24 h         │
└──────────────────────────────────────┘
                   │
                   ▼
┌──────────────────────────────────────┐
│  Dried at 110°C for 24 h             │
└──────────────────────────────────────┘
                   │
                   ▼
┌──────────────────────────────────────┐
│  Calcination at 550°C for 5 h        │
└──────────────────────────────────────┘
                   │
                   ▼
┌──────────────────────────────────────┐
│  NS–TiO₂ supported on SiO₂           │
└──────────────────────────────────────┘
```

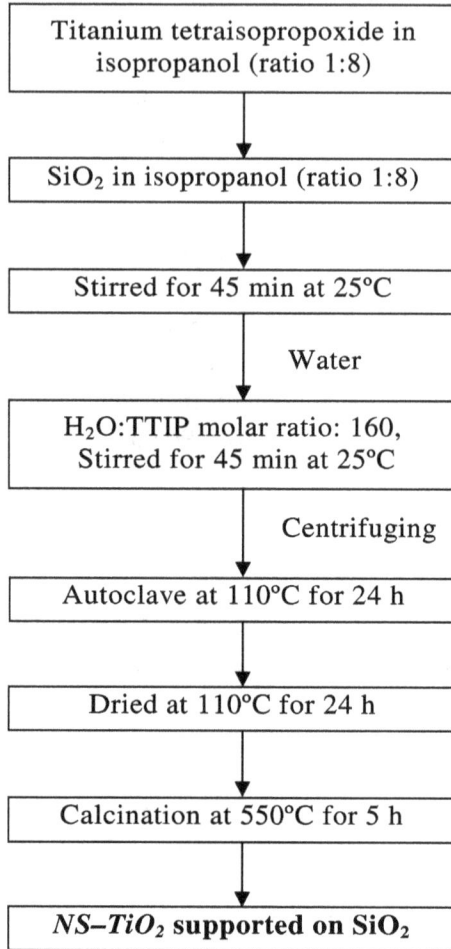

Figure 27. The synthesis procedure of SiO_2–supported *NS–TiO₂* photocatalyst.

TiO_2 nanoparticles (P25) are also supported on alumina beads by the heat attachment method. Alumina beads are initially washed with dilute nitric acid followed by distilled water. TiO_2 slurry is prepared with known amount of TiO_2 (2 g) in 100 mL distilled water and stirred overnight before introducing the support. The mixture of alumina and TiO_2 slurry is thoroughly mixed for 20 min. The mixture is then placed in an oven for 24 h at 120°C. The sample is thoroughly washed with distilled water for the removal of free TiO_2 particles [564].

Table 16. Deposition of *NS–TiO₂* on metal oxide materials.

Support substrate	Immobilization method	Objective of work	Ref.
SiO_2	Sol–gel method (see the text)	Photocatalytic degradation of dichloroacetic acid, potassium cyanide, methanol, dichloroacetic acid, *Escherichia coli*	[549–552]
Silica gel	Post–synthesis hydrolytic restructuring method	Photocatalytic degradation of methylene blue	[553]
Silica gel	Acid–catalyzed sol–gel method	Photocatalytic degradation acid orange 7	[554]
Fe_2O_3 thin films on glass	Sol–gel method	Photocatalytic inactivation of *Escherichia coli* bacteria under visible light radiation	[555]
SiO_2/Fe_3O_4	Dissolving 2.7 mL of Ti(OC₄H₉)₄ in 60 mL ethanol and adding this solution dropwise into a flask containing 800 mg of SiO₂/Fe₃O₄ support in an ethanol/water mixture (pH=5). The solution is stirred and refluxed at 90°C for 2 h. The sample is dried at 60°C and calcined at 450°C for 2 h	Preparation of magnetically separable *TiO₂* catalyst for photocatalytic degradation of Procion red MX–5B dye	[556]
$CoFe_2O_4$	Sol–gel process in reverse microemulsion combining with solvent–thermal technique	Preparation of TiO₂/CoFe₂O₄ nanomagnets photocatalysts for photocatalytic degradation of methylene blue	[557]
Fe_3O_4	Fe₃O₄:*TiO₂* core–shell prepared using homogeneous precipitation method, which takes prepared Fe₃O₄ as seeds, Ti(SO₄)₂ as raw material and CO(NH₂)₂ as precipitation reagent	Preparation of Fe₃O₄:*TiO₂* magnetic photocatalyst for photocatalytic treatment of cervical carcinoma HeLa cells	[558]
$MnFe_2O_4$	Titania is coated on MnFe₂O₄ nanoparticles by hydrolysis of titanium butoxide precursor	Preparation of magnetic and microwave absorbing of core–shell structured MnFe₂O₄/*TiO₂* nanocomposites	[559]

Table 17. Deposition of *NS–TiO₂* on metallic and metal oxide materials.

Support substrate	Immobilization method	Objective of work	Ref.
Aluminium plate	Electrophoretic deposition	Photocatalytic degradation of phenol	[546]
Foamed aluminum	Plasma spraying	Photocatalytic degradation gas–phase benzene	[560]
Metallic chromium	Radio frequency (RF) magnetron sputtering	Development a colored mirror with hydrophilicity	[561]
γ–Al_2O_3	Sol–gel and metal organic chemical vapor deposition (MOCVD)	Photocatalytic decolorization of methyl orange in aqueous medium	[562, 75]
γ–Al_2O_3	Chemical Vapor Deposition – Molecular Layering (CVD–ML)	Photocatalytic oxidation of trichloroethane and toluene in gaseous phase	[563]
Alumina beads	Heat attachment	Photocatalytic decomposition of leather dye, acid brown 14 in aqueous solution	[564]
Alumina	Sol–gel	Photocatalytic oxidation of NO_x (NO + NO_2) in gaseous phase	[565]
Cu–35Zn alloy	Arc ion plating	Preparation of photocatalytic antimicrobial coating with satisfactory mechanical property	[566]
NiTi surgical alloy	Sol–gel	Improvement of the corrosion resistance and blood compatibility of NiTi surgical alloy	[567]
Ti–6Al–7Nb alloy	Electrophoretic deposition (EPD)	Development of TiO_2 coatings on Ti–6Al–7Nb alloy	[568]
NiMnCo alloy	RF magnetic sputtering	Preparation of TiO_2 films on NiMnCo alloy and investigation of the effect of total pressure on the structure of TiO_2 films	[569]

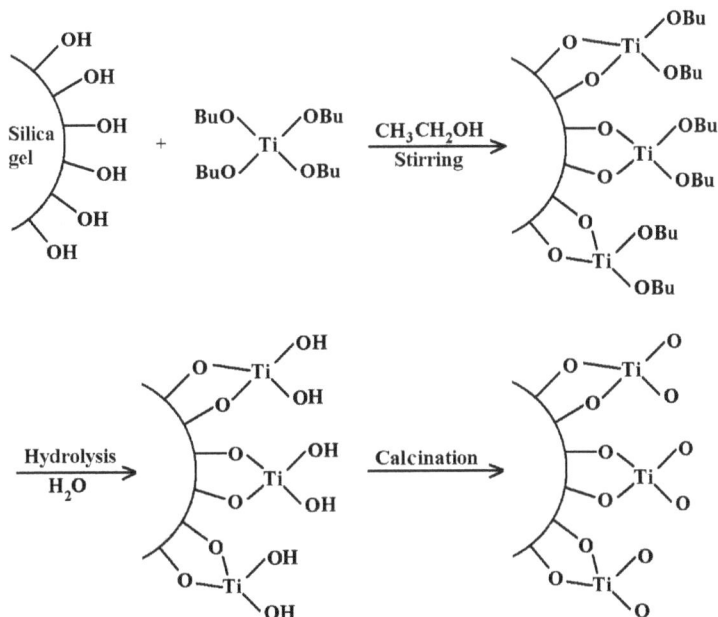

Figure 28. Schematic procedure of immobilizing $NS-TiO_2$ on silica gel, (Bu:–C_4H_9).

5.4. Immobilization on Polymer Substrates

As has already been mentioned, for practical use of $NS-TiO_2$ based photocatalytic processes, TiO_2 nanoparticles must be immobilized on a suitable substrate, because the separation of $NS-TiO_2$ from the reaction medium is difficult and costly. Hence, many techniques have been developed for immobilizing TiO_2 nanomaterials mainly on inorganic substrates such as glass, perlite, pumic stone, porous Lava, zeolites, cements pellets, ceramic tiles, metallic materials (e.g. stainless steel, aluminium plate, Foamed aluminum, chromium and alloys) and metal oxide materials (e.g. alumina, Silica and Fe_3O_4) (see sections 5.1–5.3). These substrates are stable against active oxygen species (e.g. ˙OH and $O_2^{-˙}$), which are produced through photocatalytic process.

However, $NS-TiO_2$ coating on polymer substrates is still in its early stages of development compared to the other substrates. Polymers, indeed, have many benefits as substrates for photocatalytic coating including their flexibility, low weight, impact resistance, low cost and high malleability [570]. Thus, the coated polymers with photocatalytic properties can be widely used in a diverse areas ranging from the construction or automotive industries to food packaging, kitchen utensils or toys. However, due to relatively low thermal stability of polymer substrates, the photocatalytic TiO_2 coating should be carried out with caution.

In this section, we explain different techniques for immobilizing TiO_2 nanomaterials on various polymer substrates (see Table 18).

Sol–gel has been known as the main deposition technique of NS–TiO_2 thin films for low temperature preparation on polymer substrates [571–575]. For example, Zhou *et al.* [571] reported the preparation of TiO_2–SiO_2 film with high photocatalytic activity on modified PET (Poly Ethylene Terephthalate) substrate through sol–gel method. To modify the surface of PET substrate, the cleaned PET substrate was immersed 5 min to the treatment solution which had been prepared by aminopropyl triethoxysilane dissolved in acetone (1 vol.%) and aged for 240 h. Then, the surface–modified PET substrate was formed. To prepare TiO_2–SiO_2 sol, a 0.3 M aqueous peroxotitanium acid (PTA) gel was prepared by adding 100 mL 30% H_2O_2 to 0.03 mol $TiO(OH)_2$ gel after stirring for 3 h. It was mixed with 7 mL SiO_2 sol to bring about different concentrations. The mixed sol was aged at 100°C for 9 h and the final sol was transparent and uniform. The thin films were prepared by dip–coating method, using titanium dioxide sol prepared beforehand. A dip–withdrawal rate of 3 mm/s was used for all samples. The dip–coating procedure above was repeated once and the thickness of the films ranged from 280 to 310 nm. The photocatalytic activity of TiO_2–SiO_2 thin films with different SiO_2 content was evaluated by the decomposition of Rhodamine B. At the ratio of Ti/Si=1:1, the TiO_2–SiO_2 thin film exhibited the highest photocatalytic activity due to the improved ability of the surface adsorption and the increased surface hydroxyl group in the composite thin film.

NS–TiO_2 is fixed on low–density polyethylene film (LDPE), which consists of highly–branched low crystalline units with $H(CH_2CH_2)_nH$ formula [575, 576]. The single bond presence makes this material resistance against chemical and/or corrosive agents. Polyethylene film is a flexible semitransparent low–cost commercial product. It has remarkable electrical properties, making it widely used as an insulator [575]. A 10 g/L TiO_2 Degussa P25 solution is added to 200 mL isopropanol and aged for a day to prepare LDPE–TiO_2 films. The polyethylene film is then introduced into this suspension for 10 h and the film is dried afterwards in the air at room temperature (23°C). The dried film is then heated in an oven at 180°C for 10 h to diffuse or entrain *NS–TiO_2* into the polyethylene film. Finally, LDPE–TiO_2 is sonicated for 15 min and washed with distilled water four consecutive times to eliminate the loosely bound TiO_2 nanoparticles from the film surface. The pieces of LDPE do not deform at 180°C and no pyrolysis of LDPE is observed. Complete decolorization of 0.05 mm Orange II solution has been obtained after 10 h in the presence of LDPE–TiO_2 catalyst under visible light radiation (100 mW/cm^2). LDPE–TiO_2 catalyst shows photostability over long operational periods during photocatalytic decolorization of Orange II [575].

NS–TiO_2 thin films are prepared on polymer substrates by plasma enhanced–atomic layer deposition (PE–ALD) method [570, 578]. Generally, in the ALD method, substrate temperature exceeding 300°C is needed to crystallize TiO_2 thin films. ALD has several advantages over other deposition methods such as good thin film quality, accurate thickness control and uniformity over large areas. PE–ALD has more advantages than thermal ALD such as improved film properties, high growth rate and the possibility of deposition at low temperatures. Thus, a suitable PE–ALD process practically solves the high temperature degradation problem of polymer substrates and may be a very effective method for the preparation of TiO_2 thin films on polymer substrates [578–580]. Lee *et al.* [570] prepared photocatalytic TiO_2 thin films on polyimide sheet by PE–ALD. The researchers analyzed self–cleaning effects of TiO_2/polyimide films using a contact angle measurement and photodegradation of 4–chlorophenol. The results confirmed that TiO_2/polyimide films showed superhydrophilic surfaces and decomposed 4–chlorophenol solution under UV irradiation.

124 *Nanostructured Titanium Dioxide Materials*

Table 18. Immobilization of *NS–TiO$_2$* on polymer substrates.

Polymer substrate	Immobilization method	Objective of work	Ref.
Poly ethylene terephthalate (PET)	Sol–gel and dip–coating	Photocatalytic decomposition of Rhodamine B and trichloroethylene.	[571, 572]
PET plate	Plasma spraying	Photocatalytic oxidation of acetaldehyde and toluene in gaseous phase	[573]
PET bottles	Shaking 115 mL of a *TiO$_2$* suspension at pH 2.5 (HClO$_4$) for about 30 s in transparent 330 mL PET bottles. After removing the excess suspension, the bottles are oven–dried for 20 min at 55°C to promote *TiO$_2$* adhesion. After cooling, internal washing is performed with distilled water in order to remove unfixed *TiO$_2$*	Immobilization of *TiO$_2$* on the internal surface of PET bottles; Photocatalytic oxidation of As(III) to As(V) and As(V) removal from water	[574]
Foamed polyethylene sheet	Thermal bonding	Produce a stable catalyst sheet containing 0.7 mg *TiO$_2$*/cm^2 and retaining 40–50% of active surface area of the particles; Photocatalytic decomposition of methylene blue	[576]
Polydimethylsiloxane (PDMS)	*TiO$_2$* (P25) is added at 10% (w/w) to PDMS precursor mixture and is stirred to become homogeneous. The mixture is sonicated for 1 min and the film deposited by spin–coating at 650 rpm on 64 cm^2 Petri dishes. After 36 h, it is dried in an oven at 60°C for 1 h	Photocatalytic degradation of dichloroacetic acid, salicylic acid, phenol in aqueous–phase and ethanol in gas–phase	[577]
Orthophthalic polyester (OP)	*TiO$_2$* (P25) is added at 20% (w/w) to OP precursor mixture and is stirred to become homogeneous. The film is deposited onto 27 cm^2 plastic plates using spin–coating at 850 rpm for the aqueous–phase. After 48 h, it is dried in an oven at 60°C for 1 h	Photocatalytic degradation of dichloroacetic acid, salicylic acid, phenol in aqueous–phase and ethanol in gas–phase	[577]

Table 18 (continued).

Polymer substrate	Immobilization method	Objective of work	Ref.
Polyimide sheet	Plasma enhanced atomic layer deposition (PE–ALD)	Photocatalytic degradation of 4–chlorophenol	[578]
Acrylonitrile–butadiene–styrene (ABS) and polystyrene (PS)	Dip-coating	Photocatalytic degradation of methylene blue	[581]
Buoyant polystyrene beads	Thermal treatment procedure	Photocatalytic degradation of methylene blue	[582]
Polyvinylidene fluoride (PVDF)	Magnetron sputtering	Photocatalytic degradation of RED 41 dye	[583]
Polyvinyl chloride (PVC)	Ti(OH)$_4$ sol is prepared by adding 20 mL of TiCl$_4$ to 40 mL ethanol (95%) under stirring condition at 0°C. 3.0 g PVC powders is added into the 10 mL of Ti(OH)$_4$ sol and the mixture is stirred for another 72 h. The product is filtered and dried at room temperature. After calcination at appreciate temperature for 0.5 h, *TiO$_2$*/conjugated polymer complex is obtained	Photocatalytic degradation of methylene blue	[584]
Nafion membrane	Thermal evaporation	Preparation of thin film Pt/*TiO$_2$* catalyst for t polymer electrolyte fuel cell	[585]
Polycarbonate sheet	Magnetron sputtering and a chemical surface treatment methods	Development of *TiO$_2$* self cleaning coatings for photocatalytic degradation of methylene blue and rhodamine–B	[586, 587]
Polymethyl methacrylate	Sol–gel	Photocatalytic degradation of methylene blue and acetaldehyde	[588]

As it was explained in section 4.9.3, it is of paramount importance to improve the photocatalytic efficiency of $NS-TiO_2$ by shifting its optical response from UV to the visible range, without decreasing photocatalytic activity. Doping of $NS-TiO_2$ with metal ions and non–metal atoms is the main approach to use solar radiation efficiently to conduct photocatalysis. Conductive polymers (e.g. polyaniline, polypyrrole and polythiophene) can also be used as photosensitizers to modify $NS-TiO_2$. Recently, the conductive polymers with extending π–conjugated electron systems have shown great promises thanks to their high absorption coefficients of visible light, high mobility of charge carriers and considerable stability [589, 590]. In general, they are also efficient electron donors and good hole transporters upon visible light excitation.

When a conjugated polymer/$NS-TiO_2$ system is illuminated with visible light, both $NS-TiO_2$ and conductive polymer absorb the photons at their interface, followed by the charge separation at the interface. This is because the conduction band of $NS-TiO_2$ and the lowest unoccupied molecular orbital level of the conductive polymer are well matched for the charge transfer. The electrons (e^-) generated by conductive polymer can be transferred to the conduction band of $NS-TiO_2$, enhancing the charge separation and in turn promoting the photocatalytic activity. Simultaneously, a positive–charged hole (h^+) might be formed by electrons migrating from $NS-TiO_2$ valence band to the conductive polymer. It can be concluded that the role of conductive polymer is the injection of electrons into $NS-TiO_2$ conduction band under visible light radiation. The electrons react with the electron acceptors such as oxygen or hydrogen peroxide to form very reactive radicals (i.e. $^\bullet O_2{}^-$ and $^\bullet OH$), which are responsible for degradation of organic pollutants [590–592]. A schematic diagram for the charge transfer processes of conjugated polymer and $NS-TiO_2$ is illustrated in Figure 29.

Photocatalytic applications of $NS-TiO_2$ immobilized on different conductive polymers have been summarized in Table 19.

Table 19. Immobilization of NS–TiO_2 on conductive polymers for photocatalytic applications.

Polymer substrate	Immobilization method	Removal target	Irradiation wavelength (nm)	Ref.
Polythiophene	By a two-step electrochemical process of anodization and electropolymerization	2,3–dichlorophenols	400–800	[590]
Polyaniline	*In situ* chemical oxidative polymerization method	Phenol	> 400	[591]
Polypyrrole	TiO_2 nanoparticles are suspended in 100 mL of 1.5 M HCl aqueous solution and sonicated for 30 min. Then, pyrrole is injected into the solution at 0°C with constant stirring. After that, 1.0 mL of 1.5 M HCl aqueous solution containing 0.1 g $FeCl_3$ is added dropwise. The mixture is allowed to react at 0°C for 10 h. Then, the reaction mixture is filtered, washed with 1.5 M HCl solution and large amount of deionized water respectively. Finally, the product is dried at 100°C	Methyl orange	Sunlight	[594]
Poly(3–hexylthiophene)	Chemical oxidative polymerization with anhydrous $FeCl_3$ as oxidant, 3–hexylthiophene as monomer and chloroform as solvent	Methyl orange	> 400	[595]
Polyaniline	*In situ* suspension oxidative polymerization of aniline in the presence of TiO_2 in aqueous solution	Iprobenfos fungicide	254	[596]
Polyaniline	Chemisorption approach	Methylene blue and Rhodamine B	> 450 nm	[597]
Poly(fluorene-co-thiophene) (PFT)	A mixture of 4 mg of PFT and 5 mL of tetrahydrofuran is stirred for 1 h to form a clear solution. 400 mg of TiO_2 is added to 50 mL of ethanol solution and stirred for 2 h in darkness. The PFT solution is added dropwise into the TiO_2 suspension and stirred for 10 min. The solvent is removed under vacuum. Finally, the yellow powder is dried at 60°C under vacuum. TiO_2/PFT ratio is 1:100.	Phenol	400–700	[598]

Li *et al.* [593] prepared hybrid composites of conductive polyaniline and *NS–TiO₂* through self–assembling and graft polymerization. Compared with neat *TiO₂* nanoparticles, the nanocomposites showed better photocatalytic activity in photodegradation of methyl orange under sunlight. The same results have been reported about polypyrrole–*TiO₂* nanocomposite by Wang and co–workers [594]. The researchers prepared a series of polypyrrole–*TiO₂* nanocomposites at different ratios by *in situ* deposition oxidative polymerization of pyrrole hydrochloride, using FeCl₃ as oxidant in the presence of anatase *NS–TiO₂*. Polypyrrole–*TiO₂* nanocomposites showed higher photocatalytic activity than that of neat *TiO₂* nanoparticles. This observation was partly attributed to the sensitizing effect of polypyrrole (see Figure 29). Wang *et al.* [594] found that the band gap of polypyrrole–*TiO₂* nanocomposite was smaller than that of neat *TiO₂* nanoparticles by UV–Vis diffuse reflectance spectra. The narrow band gap allows polypyrrole–*TiO₂* nanocomposite to absorb more photons that enhance the photocatalytic activity under sunlight.

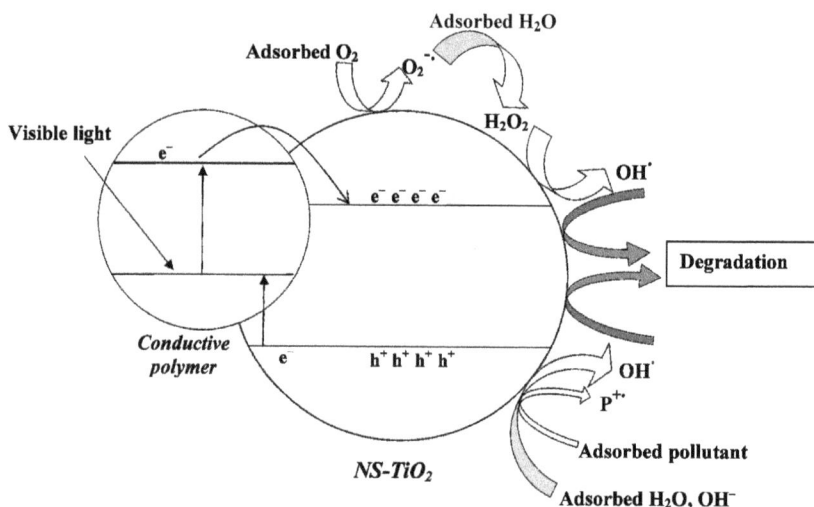

Figure 29. A schematic diagram for the charge transfer processes of conductive polymer and *NS–TiO₂*.

NS–TiO₂ has also been deposited on biopolymers such as Chitosan, Which is very popular due to its abundance, non–toxicity, hydrophilicity, biocompatibility, biodegradability and fungistatic and antibacterial activity [598–603]. A combination of adsorption and photodegradation of pollutants occurs in the presence of chitosan–TiO_2 photocatalyst. Zainal *et al.* [598] used a combination of chitosan–TiO_2 photocatalysis and adsorption processes to test the decomposition of methyl orange. In order to prepare chitosan–TiO_2 photocatalyst, 2.5 g of chitosan flake is dissolved in a premixed solution of 300 mL (0.1M) CH_3COOH and 40 mL (0.2 M) NaCl. The viscous solution is stirred continuously for 12 h to fully dissolve the chitosan flake. Then, 2.5 g of TiO_2 Degussa P25 is added into the viscous solution. Subsequently, another 50 mL of CH_3COOH is added. The slurry is stirred continuously for 24 h to obtain the final transparent viscous solution. The pieces of 45mm×80mm×2mm glass plates are used as support to immobilize the prepared chitosan–TiO_2 photocatalyst. The glass plates are first degreased, cleaned thoroughly and dried before deposition. Then, the glass plates are manually dipped in the viscous solution with a uniform pulling rate. The coated glass plates are dried at 100°C inside an oven for 4 h after each dipping process [598] (see Table 20).

Chitosan is a hydrolyzed derivative of chitin contains high amount of amino (–NH_2) and hydroxyl (–OH) functional groups. In fact, both –NH_2 and –OH groups on chitosan chains can serve as coordination and reaction sites. Adsorption of organic substrates by chitosan is via electrostatic attraction formed between –NH_2 functional groups and the solutes. Whereas, the binding ability of chitosan for metal ion is attributed to the chelating groups (–NH_2 and –OH groups) on chitosan [598, 599, 604]. The amino and hydroxyl groups on chitosan chains are good capping groups for *NS–TiO₂*. Because of its highly viscous nature, chitosan can also prevent *NS–TiO₂* from agglomeration during the growth. Also, chitosan is a well-known, excellent adsorbent for a number of organic molecules, which can further increase the photocatalytic activity of chitosan–TiO_2 catalyst.

Kim *et al.* [605] suggested a simple coating method of TiO_2 onto chitosan beads. Flaked chitosan is milled, to be able to pass through a

180–μm sieve. It is dissolved in a 2 wt% acetic acid solution to produce a viscous solution with 2 wt% chitosan. 5 g TiO_2 powders are dispersed into 100 mL of the chitosan solution during 24 h under continuous stirring. Thereafter, the mixed solution is cast into beads by a phase-inversion technique, using 2 M NaOH. The specific surface area, pore size and pore volume of TiO_2 powders are 12.02 m^2/g, 57.48 nm and 0.0172 cm^3/pore, respectively. Figure 30 shows a scheme of the manufacturing process of chitosan beads coated with TiO_2. The prepared chitosan–TiO_2 beads were used to photocatalytic disinfection of a solution containing *Salmonella choleraesuis* subsp. bacteria.

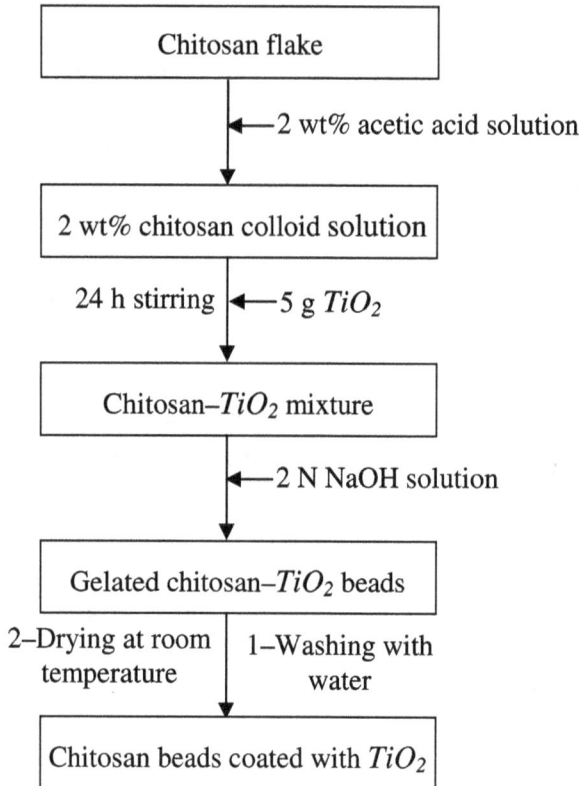

```
            ┌─────────────────────────────┐
            │       Chitosan flake        │
            └─────────────────────────────┘
                        │
                        │◄── 2 wt% acetic acid solution
                        ▼
            ┌─────────────────────────────┐
            │ 2 wt% chitosan colloid solution │
            └─────────────────────────────┘
                        │
         24 h stirring ◄── 5 g TiO₂
                        ▼
            ┌─────────────────────────────┐
            │    Chitosan–TiO₂ mixture    │
            └─────────────────────────────┘
                        │
                        │◄── 2 N NaOH solution
                        ▼
            ┌─────────────────────────────┐
            │  Gelated chitosan–TiO₂ beads │
            └─────────────────────────────┘
     2–Drying at room │ 1–Washing with
        temperature   │    water
                        ▼
            ┌─────────────────────────────┐
            │ Chitosan beads coated with TiO₂ │
            └─────────────────────────────┘
```

Figure 30. Manufacturing diagram of chitosan bead coated with TiO_2.

Table 20. Removal of pollutants using chitosan–TiO_2 catalyst through adsorption–photocatalysis process.

System	Type of TiO_2	Crystal size (nm) of TiO_2	Chitosan/TiO_2 amount	Removal target	Initial concentration	Contact time (min)	Removal efficiency (%)	Ref.
TiO_2–chitosan/Glass	Synthesized anatase	4–18	2.5 g chitosan/2.5 g TiO_2	Methyl orange	20 mg/L	–	33.7	[598]
Chitosan/activated carbon fiber/TiO_2 membrane	Synthesized TiO_2	–	0.04 g chitosan/0.02 g ACF/0.07 g TiO_2	Dichlorophenol	10 mg/L	120	> 90	[600]
TiO_2–chitosan	Anatase (Aldrich)	–	1 mg/mL TiO_2	*Salmonella choleraesuis* subsp. cells	10^8 cfu/mL	360	82.3	[605]
TiO_2–chitosan	Degussa P25	25–30	0.5 g chitosan/0.2 g TiO_2	Methyl orange	10 mg/L	360	90	[606]
TiO_2–chitosan	Synthesized anatase	> 85	280 mg chitosan/g TiO_2	Methylene blue	0.04 mm	250	91	[607]
TiO_2–chitosan	Synthesized anatase	> 85	46.76 mg chitosan/g TiO_2	Methylene blue	0.04 mm	250	41	[607]
N-doped TiO_2–chitosan	Synthesized anatase	–	Chitosan/TiO_2 ratio 1:2	Methyl orange	5 mg/L	240	97.16	[608]
TiO_2–chitosan on PET	Synthesized anatase	5–10	–	*E. coli* cells	5×10^5 cells/mL	30	99%	[609]
TiO_2–chitosan	Synthesized anatase (6 and 30 nm), P25 (27 nm)		Chitosan/TiO_2 ratio 1:1	Methylene blue and Orange II	0.025 mm	300	86.5 and 40	[610]

Discussion and Conclusions

TiO_2 nanomaterials provide a wide variety of possible applications due to their unique combination of physical and chemical properties. This book has dealt with a number of topics having to do with the properties, production, modification and applications of $NS-TiO_2$, emphasizing recent developments in these areas. Accompanied by the progress in the preparation of TiO_2 nanoparticles, there are new findings in the synthesis of TiO_2 nanorods, nanotubes, nanowires, nanosheets, as well as mesoporous structures. These new nanostructures demonstrate size–dependent as well as shape– and structure–dependent photocatalytic properties. TiO_2 nanomaterials continue to be highly active in photocatalytic and photovoltaic applications. They also demonstrate great potential in new applications such as prevention and treatment of cancer, sensors and hydrogen storage. $NS-TiO_2$ also plays an important role in the environment remediation.

We have endeavored to carry out a comprehensive review including all the relevant publications on the preparation, properties and applications of $NS-TiO_2$. However, the limitation of our resources and the sheer number of publications in this field may have compromised the comprehensiveness of this report. Our sincere apologies are extended to any and all authors whose works may not have been included in this report.

References

[1]   A.R. Khataee, M.B. Kasiri, "Photocatalytic degradation of organic dyes in the presence of nanostructured titanium dioxide: Influence of the chemical structure of dyes", Journal of Molecular Catalysis A: Chemical, 328, 8–26, (2010).

[2]   R. Zallen, M.P. Moret, "The optical absorption edge of brookite TiO₂", Solid State Communications, 137, 154–157, (2006).

[3]   Kirk–Othmer, "Encyclopedia of Chemical Technology", Wiley–Interscience publication, Fourth Edition, Volume 19, (1996).

[4]   A. Fujishima, T.N. Rao, D.A. Tryk, "Titanium dioxide photocatalysis", Journal of Photochemistry and Photobiology C: Photochemistry Reviews, 1, 1–21, (2000).

[5]   A. Fujishima, A.X. Zhang, "Titanium dioxide photocatalysis: present situation and future approaches", Comptes Rendus Chimie, 9, 750–760, (2006).

[6]   U.I. Gaya, A.H. Abdullah, "Heterogeneous photocatalytic degradation of organic contaminants over titanium dioxide: A review of fundamentals, progress and problems", Journal of Photochemistry and Photobiology C: Photochemistry Reviews, 9, 1–12, (2008).

[7]   V.M. Prida, E. Manova, V. Vega, M. Hernandez–Velez, P. Aranda, K.R. Pirota, M. Vazquez, E. Ruiz–Hitzky, "Temperature influence on the anodic growth of self–aligned Titanium dioxide nanotube arrays", Journal of Magnetism and Magnetic Materials, 316, 110–113, (2007).

[8]   G.A. Mansoori, P. Mohazzabi, "Why nanosystems and macroscopicsystems behave differently", International Journal of Nanoscience & Nanotechnology, 1(1), 46–53, (2006).

[9]   G.A. Mansoori, "Principles of nanotechnology–molecular–based study of condensed matter in small systems", World Scientific Publishing Company, (2005).

[10]   G.A. Mansoori, C. Darnault, K. Rockne, A. Stevens, N. Sturchio, "Nanoparticles, in Fate of environmental pollutants", Water Environment Research, 77, 2576–2658, (2005).

[11]   G.A. Mansoori, G.R. Vakili–Nezhaad, A.R. Ashrafi, "Symmetry Property of Fullerenes", Journal of Computational and Theoretical Nanoscience, 4(5), 1–4, (2007).

[12]   G.A. Mansoori, H. Ramezani, W.R. Saberi, "Diamondoids–DNA nanoarchitecture: from nanomodules design to self–assembly", Journal of Computational and Theoretical Nanoscience, 4(1), 96–106, (2007).

[13]   G.A. Mansoori, A. Nikakhtar, A. Nasehzadeh, "Formation and stability conditions of DNA–Dendrimer nano–clusters", Journal of Computational and Theoretical Nanoscience, 4(3), 521–528, (2007).

[14]   A.P. Alivisatos, "Perspectives on the physical chemistry of semiconductor nanocrystals", Journal of Physical Chemistry, 100 (31), 13226–13239, (1996).

[15]   C. Burda, X. Chen, R. Narayanan, M.A. El–Sayed, "Chemistry and properties of nanocrystals of different shapes", Chemical Reviews, 105 (4), 1025–1102, (2005).

[16]   K. Pirkanniemi, M. Sillanpaa, "Heterogeneous water phase catalysis as an environmental application: a review", Chemosphere, 48, 1047–1060, (2002).

[17]   A.R. Khataee, V. Vatanpour, A.R. Amani, "Decolorization of C.I. Acid blue 9 solution by UV/Nano–TiO$_2$, Fenton, Electro–Fenton and Electrocoagulation processes: A comparative study", Journal of Hazardous Materials, 161, 1225–1233, (2009).

[18]   R. Andreozzi, V. Caprio, A. Insola, R. Marotta, "Advanced oxidation processes (AOP) for water purification and recovery", Catalysis Today, 53, 51–59, (1999).

[19]   S. Kwon, M. Fan, A.T. Cooper, H. Yang, "Photocatalytic applications of micro– and nano–TiO$_2$ in environmental engineering", Critical Reviews in Environmental Science and Technology, 38, 197–226, (2008).

[20] I.K. Konstantinou, T.A. Albanis, "TiO$_2$–assisted photocatalytic degradation of azo dyes in aqueous solution: kinetic and mechanistic investigations A review", Applied Catalysis B: Environmental, 49, 1–14, (2004).

[21] E. Katz, I. Willner, "Integrated Nanoparticle–Biomolecule hybrid systems: Synthesis, properties and applications", Angewandte Chemie–International Edition, 43, 6042 – 6108, (2004).

[22] X. Chen, S.S. Mao, "Titanium dioxide nanomaterials: synthesis, properties, modifications and applications", Chemical Reviews, 107, 2891–2959, (2007).

[23] M.A. Fox, M.T. Dulay, "Heterogeneous photocatalysis", Chemical Reviews, 83, 341–357, (1993).

[24] T. Yates, Jr, A.L. Linsebigler, G. Lu, "Photocatalysis on TiO$_2$ surfaces: principles, mechanisms and selected results", Chemical Reviews, 95, 735–758, (1995).

[25] T. Yates, Jr, T.L. Thompson, "Surface science studies of the photoactivation of TiO$_2$ new photochemical processes", Chemical Reviews, 106, 4428–4453, (2006).

[26] M.R. Hoffmann, S.T. Martin, W. Choi, D.W. Bahnemannt, "Environmental applications of semiconductor photocatalysis", Chemical Reviews, 95, 69–96, (1995).

[27] Daniel M. Blake, "Bibliography of work on the heterogeneous photocatalytic removal of hazardous compounds from water and air", National Renewable Energy Laboratory, Update Number 4 to October 2001.

[28] C. Walsh, D.V. Bavykin, J.M. Friedrich, "Protonated Titanates and TiO$_2$ nanostructured materials: synthesis, properties and applications", Advanced Materials, 18, 2807–2824, (2006).

[29] C.A. Grimes, G.K. Mor, O.K. Varghese, M. Paulose, K. Shankar, "A review on highly ordered, vertically oriented TiO$_2$ nanotube arrays: Fabrication, material properties and solar energy applications", Solar Energy Materials & Solar Cells, 90, 2011–2075, (2006).

[30] U. Diebold, "The surface science of titanium dioxide", Surface Science Reports, 48, 53–229, (2003).

[31] Y. Li, T.J. White, S.H. Lim, "Low–temperature synthesis and microstructural control of titania nano–particles", Journal of Solid State Chemistry, 177, 1372–1381, (2004).

[32] M. Yan, F. Chen, J. Zhang, M. Anpo, "Preparation of controllable crystalline Titania and study on the photocatalytic properties", Journal of Physical Chemistry B, 109, 8673–8678, (2005).

[33] R. Janisch, P. Gopal, N. Spaldin, "Transition metal–doped TiO$_2$ and ZnO–present status of the field", Journal of Physics: Condensed Matter 17, 657–689, (2005).

[34] S. Lee, S. Byeon, "Structural and morphological behavior of TiO$_2$ rutile obtained by hydrolysis reaction of Na$_2$Ti$_3$O$_7$", Bulletin of the Korean Chemical Society, 25, 1051–1054, (2004).

[35] T.P. Feist, P.K. Davies, "The soft chemical synthesis of TiO$_2$ (B) from layered titanates", Journal of Solid State Chemistry, 101, 275–295, (1992).

[36]   A.K. Chakraborty, Z. Qi, S. Yong Chai, C. Lee, S. Park, D.J. Jang, W.I. Lee, "Formation of highly crystallized $TiO_2(B)$ and its photocatalytic behavior", Applied Catalysis B: Environmental, 93, 368–375, (2010).

[37]   A.R. Armstrong, G. Armstrong, J. Canales, R. Garcia, P.G. Bruce, "Lithium–Ion intercalation into TiO_2–B nanowires", Advanced Materials, 17, 862–865, (2005).

[38]   G. Nuspl, K. Yoshizawa, T. Yamabe, "Lithium intercalation in TiO_2 modifications", Journal of Materials Chemistry, 7, 2529–2536, (1997).

[39]   J. Zhu, J. Zhang, F. Chen, M. Anpo, "Preparation of high photocatalytic activity TiO_2 with a bicrystalline phase containing anatase and TiO_2 (B)", Materials Letters, 59, 3378–3381, (2005).

[40]   A.R. Armstrong, G. Armstrong, J. Canales, P.G. Bruce, "TiO_2–B Nanowires", Angewandte Chemie International Edition, 43, 2286–2288, (2004).

[41]   A.R. Khataee, "Photocatalytic removal of C.I. Basic Red 46 on immobilized TiO_2 nanoparticles: Artificial neural network modeling", Environmental Technology, 30, 2009, 1155–1168

[42]   N. Daneshvar, M.H. Rasoulifard, F. Hosseinzadeh, A.R. Khataee, "Removal of C.I. Acid Orange 7 from aqueous solution by UV irradiation in the presence of ZnO nanopowder", Journal of Hazardous Materials, 143, 95–101, (2007).

[43]   M.P. Moret, R. Zallen, D.P. Vijay, S.B. Desu, "Brookite–rich titania films made by pulsed laser deposition", Thin Solid Films, 366, 8–10, (2000).

[44]   G.A. Mansoori, T. Rohani. Bastami, A. Ahmadpour, Z. Eshaghi, "Environmental applications of nanotechnology", Annual Review of Nano Research, Vol. 2, Chapter 2, 1–73, (2008).
         See also/ http://ruby.colorado.edu/~smyth/min/tio2.html.

[45]   A. Mills, S. Le Hunte, "An overview of semiconductor photocatalysis" Journal of Photochemistry and Photobiology A: Chemistry, 108, 1–35, (1997).

[46]   N. Daneshvar, D. Salari, A.R. Khataee, "Photocatalytic degradation of azo dye acid red 14 in water: investigation of the effect of operational parameters", Journal of Photochemistry and Photobiology A: Chemistry, 157, 111–116, (2003).

[47]   N. Daneshvar , D. Salari, A.R. Khataee, "Photocatalytic degradation of azo dye acid red 14 in water on ZnO as an alternative catalyst to TiO_2", Journal of Photochemistry and Photobiology A: Chemistry, 162, 317–322, (2004).

[48]   X.F. Cheng, W.H. Leng, D.P. Liu, J.Q. Zhang, C.N. Cao, "Enhanced photoelectrocatalytic performance of Zn–doped WO_3 photocatalysts for nitrite ions degradation under visible light", Chemosphere, 68, 1976–1984, (2007).

[49]   D. Jing, L. Guo, "WS2 sensitized mesoporous TiO_2 for efficient photocatalytic hydrogen production from water under visible light irradiation", Catalysis Communications, 8, 795–799, (2007).

[50]   J. Bandara, U. Klehm, J. Kiwi, "Raschig rings–Fe_2O_3 composite photocatalyst activate in the degradation of 4–chlorophenol and Orange II under daylight irradiation", Applied Catalysis B: Environmental, 76, 73–81, (2007).

[51]   K. Teramura, T. Tanaka, M. Kani, T. Hosokawa, T. Funabiki, "Selective photo–oxidation of neat cyclohexane in the liquid phase over V_2O_5/Al_2O_3", Journal of Molecular Catalysis A: Chemical, 208, 299–305, (2004).

[52]   Y. Zhai, S. Zhang, H. Pang, "Preparation, characterization and photocatalytic activity of CeO_2 nanocrystalline using ammonium bicarbonate as precipitant", Materials Letters, 61,1863–1866, (2007).

[53]   K.G. Kanade, Jin–OoK Baeg, U.P. Mulik, D.P. Amalnerkar, B.B. Kale, "Nano–CdS by polymer–inorganic solid–state reaction: Visible light pristine photocatalyst for hydrogen generation", Materials Research Bulletin, 41, 2219–2225, (2006).

[54]   C.L. Torres–Martínez, R. Kho, O.I. Mian, R.K. Mehra, "Efficient photocatalytic degradation of environmental pollutants with mass–produced ZnS nanocrystals", Journal of Colloid and Interface Science, 240, 525–532, (2001).

[55]   R. Richards, "Surface and nanomolecular catalysis", Published by CRC Press, Taylor & Francis Group, (2006).

[56]   N. Daneshvar, D. Salari, A. Niaie, M.H. Rasoulifard, A.R. Khataee, "Immobilization of TiO_2 nanopowder on glass beads for the photocatalytic decolorization of an azo dye C.I. direct Red 23", Journal of Environmental Science and Health, Part A, 40, 1605–1617, (2005).

[57]   N. Daneshvar, A. Aleboyeh, A.R. Khataee, "The evaluation of electrical energy per order (E_{Eo}) for photooxidative decolorization of four textile dye solutions by the kinetic model", Chemosphere, 59, 761–767, (2005).

[58]   V. Loddo, G. Marcõ, C. Martõ, L. Palmisano,V. Rives, A. Sclafania, "Preparation and characterisation of TiO_2 (anatase) supported on TiO_2 (rutile) catalysts employed for 4–nitrophenol photodegradation in aqueous medium and comparison with TiO_2 (anatase) supported on Al_2O_3", Applied Catalysis B: Environmental, 20, 29–45, (1999).

[59]   S. Bakardjieva, J. Subrt, V. Stengl, M. Dianez, M. Sayagues, "Photoactivity of anatase–rutile TiO_2 nanocrystalline mixtures obtained by heat treatment of homogeneously precipitated anatase", Applied Catalysis B: Environmental, 58, 193–202, (2005).

[60]   S. Jung, S. Kim, N. Imaishi, Y. Cho, "Effect of TiO_2 thin film thickness and specific surface area by low–pressure metal–organic chemical vapor deposition on photocatalytic activities", Applied Catalysis B: Environmental, 55, 253–257, (2005).

[61]   K. Madhusudan Reddy, D. Guin, S.V. Manoramaa, A.R. Reddy, "Selective synthesis of nanosized TiO_2 by hydrothermal route: Characterization, structure property relation and photochemical application", Journal of Materials Research, 19, 2567–2575, (2004).

[62]   M. Nakamura, N. Negishi, S. Kutsuna, T. Ihara, S. Sugihara, K. Takeuchi, "Role of oxygen vacancy in the plasma–treated TiO_2 photocatalyst with visible light

activity for NO removal", Journal of Molecular Catalysis A: Chemistry, 161, 205–212, (2000).

[63]   N.L. Wu, M.S. Lee, Z.J. Pon, J.Z. Hsu, "Effect of calcination atmosphere on TiO_2 photocatalysis in hydrogen production from methanol/water solution", Journal of Photochemstry and Photobiology. A: Chemistry, 163, 277–280, (2004).

[64]   X. Wang, J.C. Yu, P. Liu, X. Wang, W. Sua, X. Fua, "Probing of photocatalytic surface sites on SO_4/TiO_2 solid acids by in situ FT–IR spectroscopy and pyridine adsorption", Journal of Photochemistry and Photobiology A: Chemistry, 179, 339–347, (2006).

[65]   M. Zheng, M. Gu, Y. Jin, G. Jin, "Preparation, structure and properties of TiO_2–PVP hybrid films", Materials Science and Engineering B, 77, 55–59, (2000).

[66]   N. Daneshvar, D. Salari, A. Niaei, A. R. Khataee, "Photocatalytic degradation of the herbicide Erioglaucine in the presence of nanosized titanium dioxide: comparison and modeling of reaction kinetics", Journal of Environmental Science and Health, Part B, 41, 1273–1290, (2006).

[67]   K. Tennakone, K.G.U. Wijayantha, "Heavy–metal extraction from aqueous medium with an immobilized TiO_2 photocatalyst and a solid sacrificial agent", Journal of Photochemistry and Photobiology A: Chemistry, 113, 89–92, (1998).

[68]   M.N. Pons, A. Alinsafi, F. Evenou, E.M. Abdulkarim, O. Zahraa, A. Benhammou, A. Yaacoub and A. Nejmeddine, "Treatment of textile industry wastewater by supported photocatalysis", Dyes and Pigments, 74, 439–445, (2007).

[69]   N. Keller, G. Rebmann, E. Barraud, O. Zahraa, V. Keller, "Macroscopic carbon nanofibers for use as photocatalyst support", Catalysis Today, 101, 323–329, (2005).

[70]   K. Ping Loh, S. Chua, X. Wang, "Diamondoids as molecular building blocks for nanotechnology, from diamondoids to nanoscale materials and applications ", edited by G.A. Mansoori, T.F. George, L. Assoufid and G.P. Zhang (Springer, New York), Topics in Applied Physics 109, 92–118, (2007).

[71]   K. Mori, K. Maki, S. Kawasaki, S. Yuan, H. Yamashita, "Hydrothermal synthesis of TiO_2 photocatalysts in the presence of NH4F and their application for degradation of organic compounds", Chemical Engineering Science, 63, 5066–5070, (2008).

[72]   Z. Li, B. Houa, Y. Xu, D. Wu, Y. Sun, "Hydrothermal synthesis, characterization and photocatalytic performance of silica–modified titanium dioxide nanoparticles", Journal of Colloid and Interface Science, 288, 149–154, (2005).

[73]   L. Kao, T. Hsu, H. Lu, "Sol–gel synthesis and morphological control of nanocrystalline TiO_2 via urea treatment", Journal of Colloid and Interface Science, 316, 160–167, (2007).

[74]   S. Ansari Amin, M. Pazouki, A. Hosseinnia, "Synthesis of TiO_2–Ag nanocomposite with sol–gel method and investigation of its antibacterial activity against *E. coli*", Powder Technology, 196, 241–245, (2009).

[75]   X. Zhang, M. Zhou, L. Lei, "Preparation of anatase TiO_2 supported on alumina by different metal organic chemical vapor deposition methods", Applied Catalysis A: General, 282, 285–293, (2005).

[76]   H. Yoshitake, T. Sugihara, T. Tatsumi, "Preparation of wormhole–like mesoporous TiO_2 with an extremely large surface area and stabilization of its surface by chemical vapor deposition", Chemistry of Materials, 14(3), 1023–1029, (2002).

[77]   D. Byun, Y. Jin, B. Kim, J. Kee Lee, D. Park "Photocatalytic TiO_2 deposition by chemical vapor deposition", Journal of Hazardous Materials, 73, 199-206, (2000).

[78]   C. Giolli, F. Borgioli, A. Credi, A. Fabio, A. Fossati, M. Miranda, S. Parmeggiani, G. Rizzi, A. Scrivani, S. Troglio, "Characterization of TiO_2 coatings prepared by a modified electric arc–physical vapour deposition system", Surface and Coatings Technology, 202,13–22, (2007).

[79]   S. Chiu, Z. Chen, K. Yang, Y. Hsu, D. Gan, "Photocatalytic activity of doped TiO_2 coatings prepared by sputtering deposition", Journal of Materials Processing Technology, 192–193, 60–67, (2007).

[80]   M. Kang, "The superhydrophilicity of Al–TiO_2 nanometer sized material synthesized using a solvothermal method", Materials Letters, 59, 3122–3127, (2005).

[81]   R.K. Wahi, Y. Liu, J.C. Falkner, V.L. Colvin, "Solvothermal synthesis and characterization of anatase TiO_2 nanocrystals with ultrahigh surface area", Journal of Colloid and Interface Science, 302, 530–536, (2006).

[82]   B.R. Sankapal, S.D. Sartale, M.C. Lux–Steiner, A. Ennaoui, "Chemical and electrochemical synthesis of nanosized TiO_2 anatase for large–area photon conversion", Comptes Rendus Chimie, 9, 702–707, (2006).

[83]   H.E. Prakasam, K. Shankar, M. Paulose, O.K. arghese, C.A. Grimes, "A new benchmark for TiO_2 nanotube array growth by anodization", Journal of Physical Chemistry C, 111(20), 7235–7241, (2007).

[84]   N.R. Tacconi, C.R. Chenthamarakshan, G. Yogeeswaran, A. Watcharenwong, R.S. de Zoysa, N.A. Basit, K. Rajeshwar, "Nanoporous TiO_2 and WO_3 films by anodization of titanium and tungsten substrates: influence of process variables on morphology and photoelectrochemical response", Journal of Physical Chemistry B, 110(50), 25347–25355, (2006).

[85]   K. Nagaveni, G. Sivalingam, M.S. Hegde, G. Madras, "Photocatalytic degradation of organic compounds over combustion–synthesized nano–TiO_2", Environmental Science and Technology, 38(5), 1600–1604, (2004).

[86]   G. Sivalingam, M.H. Priya, G. Madras, "Kinetics of the photodegradation of substituted phenols by solution combustion synthesized TiO_2", Applied Catalysis B: Environmental, 51, 67–76, (2004).

[87]   G. Sivalingam, G. Madras, "Photocatalytic degradation of poly(bisphenol–A–carbonate) in solution over combustion–synthesized TiO_2: mechanism and kinetics", Applied Catalysis A: General, 269, 81–90, (2004).

[88]   T. Mishra, "Anion supported TiO_2–ZrO_2 nanomaterial synthesized by reverse microemulsion technique as an efficient catalyst for solvent free nitration of halobenzene", Catalysis Communications, 9, 21–26, (2008).

[89]   M. Lee, G. Lee, C. Ju, S. Hong, "Preparations of nanosized TiO_2 in reverse microemulsion and their photocatalytic activity", Solar Energy Materials and Solar Cells, 88, 389–401, (2005).

[90]   S. Priyanto, G.A. Mansoori, A. Suwono, "Measurement of property relationships of nano–structure micelles and coacervates of asphaltene in a pure solvent", Chemical Engineering Science, 56, 6933–6939, (2001).

[91]   X. Sui, Y. Chu, S. Xing, M. Yu, C. Liu, "Self–organization of spherical PANI/TiO_2 nanocomposites in reverse micelles", Colloids and Surfaces A: Physicochemical and Engineering Aspects, 251, 103–107, (2004).

[92]   B.K. Kim, G.G. Lee, H.M. Park, N.J. Kim, "Characteristics of nanostructured TiO_2 powders synthesized by combustion flame–chemical vapor condensation process", Nanostructured Materials, 12, 637–640, (1999).

[93]   N.G. Glumac, Y.J. Chen, G. Skandan, B. Kear, "Scalable high–rate production of non–agglomerated nanopowders in low pressure flames", Materials Letters, 34, 148–153, (1998).

[94]   W. Guo, Z. Lin, X. Wang, G. Song, "Sonochemical synthesis of nanocrystalline TiO_2 by hydrolysis of titanium alkoxides", Microelectronic Engineering, 66, 95–101, (2003).

[95]   T. Miyata, S. Tsukada, T. Minami, "Preparation of anatase TiO_2 thin films by vacuum arc plasma evaporation", Thin Solid Films, 496, 136–140, (2006).

[96]   H. Huang, X. Yao, "Preparation and characterization of rutile TiO_2 thin films by mist plasma evaporation", Surface and Coatings Technology, 191, 54–58, (2005).

[97]   H. Huang, X. Yao, "Preparation of rutile TiO_2 thin films by mist plasma evaporation", Journal of Crystal Growth, 268, 564–567, (2004).

[98]   P. Knauth, J. Schoonman, "Nanostructured materials: selected synthesis methods, properties and applications", Kluwer Academic Publishers, 2004.

[99]   G. Wang, "Hydrothermal synthesis and photocatalytic activity of nanocrystalline TiO_2 powders in ethanol–water mixed solutions", Journal of Molecular Catalysis A: Chemical, 274, 185–191, (2007).

[100] D. Suk Kim, S. Kwak, "The hydrothermal synthesis of mesoporous TiO_2 with high crystallinity, thermal stability, large surface area and enhanced photocatalytic activity", Applied Catalysis A: General, 323, 110–118, (2007).

[101] C. Colbeau–Justin, Y.V. Kolen'ko, B.R. Churagulov, M. Kunst, L. Mazerolles, " Photocatalytic properties of titania powders prepared by hydrothermal method", Applied Catalysis B: Environmental 54, 51–58, (2004).

[102] B. Jiang, H. Yin, T. Jiang, Y. Jiang, H. Feng, K. Chen, W. Zhou, Y. Wada, "Hydrothermal synthesis of rutile TiO_2 nanoparticles using hydroxyl and carboxyl group–containing organics as modifiers", Materials Chemistry and Physics, 98, 231–235, (2006).

[103] S. Chae, M. Park, S. Lee, T. Kim, S. Kim, W. Lee, "Preparation of size–controlled TiO$_2$ nanoparticles and derivation of optically transparent photocatalytic films", Chemistry of Materials, 15, 3326–3331, (2003).

[104] S. V. Manorama, M. Nag, P. Basak, "Low–temperature hydrothermal synthesis of phase–pure rutile titania nanocrystals: Time temperature tuning of morphology and photocatalytic activity", Materials Research Bulletin, 42, 1691–1704, (2007).

[105] S. Yoshikawa, J. Jitputti, S. Pavasupree, Y. Suzuki, "Synthesis and photocatalytic activity for water–splitting reaction of nanocrystalline mesoporous titania prepared by hydrothermal method", Journal of Solid State Chemistry, 180, 1743–1749, (2007).

[106] S. Yoshikawa, S. Pavasupree, J. Jitputti, S. Ngamsinlapasathian, "Hydrothermal synthesis, characterization, photocatalytic activity and dye–sensitized solar cell performance of mesoporous anatase TiO$_2$ nanopowders", Materials Research Bulletin, 43, 149–157, (2008).

[107] J. Yu, G. Wang, B. Cheng, M. Zhou, "Effects of hydrothermal temperature and time on the photocatalytic activity and microstructures of bimodal mesoporous TiO$_2$ powders", Applied Catalysis B: Environmental, 69, 171–180, (2007).

[108] C. Tian, Z. Zhang, J. Hou, N. Luo, "Surfactant/co–polymer template hydrothermal synthesis of thermally stable, mesoporous TiO$_2$ from TiOSO$_4$", Materials Letters, 62, 77–80, (2008).

[109] J. Yu, H. Yu, B. Cheng, X. Zhao, Q. Zhang, "Preparation and photocatalytic activity of mesoporous anatase TiO$_2$ nanofibers by a hydrothermal method", Journal of Photochemistry and Photobiology A: Chemistry, 182, 121–127, (2006).

[110] Y. Ma, Y. Lin, X. Xiao, X. Zhou, X. Li, "Sonication–hydrothermal combination technique for the synthesis of titanate nanotubes from commercially available precursors", Materials Research Bulletin, 41, 237–243, (2006).

[111] O. Yang, M.A. Khan, H. Jung, "Synthesis and characterization of ultrahigh crystalline TiO$_2$ nanotubes", Journal of Physical Chemistry B, 110, 6626–6630, (2006).

[112] J. Yu, H. Yu, "Facile synthesis and characterization of novel nanocomposites of titanate nanotubes and rutile nanocrystals", Materials Chemistry and Physics, 100, 507–512, (2006).

[113] D. Weng, W. Chen, X. Sun, "Morphology control of titanium oxides by tetramethylammonium cations in hydrothermal conditions", Materials Letters, 60, 3477–3480, (2006).

[114] H. Ou, S. Lo, "Review of titania nanotubes synthesized via the hydrothermal treatment: Fabrication, modification and application", Separation and Purification Technology, 58, 179–191, (2007).

[115] B. Chi, E. Victorio, T. Jin, "Synthesis of Eu–doped photoluminescent titania nanotubes via a two–step hydrothermal treatment", Nanotechnology, 17, 2234–2241, (2006).

[116] B. Chi, E. Victorio, T. Jin, "Synthesis of TiO$_2$–based nanotube on Ti substrate by hydrothermal treatment", Journal of Nanoscience and Nanotechnology, 7(2), 668–672, (2007).

[117] G. Armstrong, A.R. Armstrong, J. Canales, P.G. Bruce, "Nanotubes with the TiO$_2$–B structure", Chemical Communications, 21, 2454–2456, (2005).

[118] M. Qamar, C. Yoon, H . Oh, D. Kim, J. Jho, K. Lee, W. Lee, H. Lee, S. Kim, "Effect of post treatments on the structure and thermal stability of titanate nanotubes", Nanotechnology, 17, 5922–5929, (2006).

[119] M.R. Kim , S. Ahn , D. Jang, "Preparation and characterization of titania/ZnS core–shell nanotubes", Journal of Nanoscience and Nanotechnology, 6(1), 180–184, (2006).

[120] X. Ding, X .G Xu, Q. Chen, L–M. Peng, "Preparation and characterization of Fe–incorporated titanate nanotubes", Nanotechnology, 17, 5423–5427, (2006).

[121] R. Yoshida, Y. Suzuki, S. Yoshikawa, "Syntheses of TiO$_2$(B) nanowires and TiO$_2$ anatase nanowires by hydrothermal and post–heat treatments", Journal of Solid State Chemistry, 178 (7), 2179–2185, (2005).

[122] S. Pavasupree, S. Ngamsinlapasathian, Y. Suzuki, S. Yoshikawa, "Preparation and characterization of high surface area nanosheet titania with mesoporous structure" Materials Letters, 61, 2973–2977, (2007).

[123] Y. Chena, C. Lee, M. Yeng, H. Chiu, "Preparing titanium oxide with various morphologies", Materials Chemistry and Physics, 81, 39–44, (2003).

[124] Z. Song, H. Xu, K. Li, H. Wang, H. Yan, "Hydrothermal synthesis and photocatalytic properties of titanium acid H$_2$Ti$_2$O$_5$·H$_2$O nanosheets", Journal of Molecular Catalysis A: Chemical, 239, 87–91, (2005).

[125] R. Yoshida, Y. Suzuki, S. Yoshikawa, "Effects of synthetic conditions and heat–treatment on the structure of partially ion–exchanged titanate nanotubes", Materials Chemistry and Physics, 91, 409–416, (2005).

[126] H. Li, F. Zhang, "Hydrothermal synthesis of TiO$_2$ nanofibers" Materials Science and Engineering, 27, 80–82, (2007).

[127] Y. Suzuki, S. Yoshikawa, "Rapid communications: synthesis and thermal analyses of TiO$_2$–derived nanotubes prepared by the hydrothermal method", Journal of Materials Research, 19, 982–985, (2004).

[128] S. Pavasupree, Y. Suzuki, S. Yoshikawa, R. Kawahata, "Synthesis of titanate, TiO$_2$ (B) and anatase TiO$_2$ nanofibers from natural rutile sand", Journal of Solid State Chemistry, 178, 3110–3116, (2005).

[129] Y. Suzuki, S. Pavasupree, S. Yoshikawa, R. Kawahata, "Natural rutile–derived titanate nanofibers prepared by direct hydrothermal processing", Journal of Materials Research., 20, 1063–1070, (2005).

[130] L.L. Hench, J.K. West, "The sol–gel process", Chemical Reviews, 90. 33–72, (1990).

[131] P.J. Flory, L.L. Hench, D.R. Ulrich, "Science of Ceramic Chemical Processing", Wiley: New York, (1986).

[132] H. WC, F. SH, T. JJ, C. H, K. TH, "Study on photocatalytic degradation of gaseous dichloromethane using pure and iron ion–doped TiO$_2$ prepared by the sol–gel method", Chemosphere, 66 (11), 2142–2151, (2007).

[133] P.J. Flory, "Principles of polymer chemistry", Cornell University, Press: Ithaca, NY, Chapter IX, (1953).

[134] Q. Xiao, Z. Si, Z. Yu, G. Qiu, "Sol–gel auto–combustion synthesis of samarium-doped TiO$_2$ nanoparticles and their photocatalytic activity under visible light irradiation", Materials Science and Engineering: B, 137, 189–194, (2007).

[135] H. Choi, E. Stathatos, D. Dionysiou, "Synthesis of nanocrystalline photocatalytic TiO$_2$ thin films and particles using sol–gel method modified with nonionic surfactants" , Thin Solid Films, 510, 107–114, (2006).

[136] N. Venkatachalam, M. Palanichamy, V. Murugesan, "Sol–gel preparation and characterization of alkaline earth metal doped nano TiO$_2$: Efficient photocatalytic degradation of 4–chlorophenol", Journal of Molecular Catalysis A: Chemical, 273, 177–185,(2007).

[137] M. Sig Lee, S. Hong, M. Mohseni, "Synthesis of photocatalytic nanosized TiO$_2$–Ag particles with sol–gel method using reduction agent", Journal of Molecular Catalysis A: Chemical, 242, 135–140, (2005).

[138] X. Li, J. Gao, Z. Zeng, "pH–resistant titania hybrid organic–inorganic sol–gel coating for solid–phase microextraction of polar compounds", Analytica Chimica Acta, 590 (1), 26–33, (2007).

[139] V. Murugesan, N. Venkatachalam, M. Palanichamy, "Sol–gel preparation and characterization of nanosize TiO$_2$: Its photocatalytic performance", Materials Chemistry and Physics 104, 454–459, (2007).

[140] B. Li, X. Wang, M. Yan, L. Li, "Preparation and characterization of nano–TiO$_2$ powder", Materials Chemistry and Physics, 78, 184–188, (2002).

[141] A.R. Phani, D. Claudio, S. Santucci, "Enhanced optical properties of sol–gel derived TiO$_2$ films using microwave irradiation", Optical Materials, 30, 279–284, (2007).

[142] K.K. Saini, Sunil Dutta Sharma, Chanderkant, Meenakshi Kar, Davinder Singh, C.P. Sharma, "Structural and optical properties of TiO$_2$ thin films derived by sol–gel dip coating process", Journal of Non–Crystalline Solids, 353, 2469–2473, (2007).

[143] P.B. Kuyyadi, K.J. Mahaveer, "Effect of crystallization on humidity sensing properties of sol–gel derived nanocrystalline TiO$_2$ thin films", Thin Solid Films, 516, 2175–2180, (2008).

[144] M. Addamo, V. Augugliaro, A. Di Paola, E. García–López, V. Loddo, G. Marcì, L. Palmisano, "Photocatalytic thin films of TiO$_2$ formed by a sol–gel process using titanium tetraisopropoxide as the precursor", Thin Solid Films, 516, 3802–3807, (2008).

[145] R. Mechiakh, R. Bensaha, "Variation of the structural and optical properties of sol–gel TiO_2 thin films with different treatment temperatures", Comptes Rendus Physique, 7, 464–470, (2006).

[146] M. Houmard, D. Riassetto, F. Roussel, A. Bourgeois, G. Berthomé, J.C. Joud, M. Langlet, "Morphology and natural wettability properties of sol–gel derived TiO_2–SiO_2 composite thin films", Applied Surface Science, 254, 1405–1414, (2007).

[147] Z. Miao, D. Xu, J. Ouyang, G. Guo, X. Zhao, Y. Tang, "Electrochemically induced Sol–Gel preparation of single–crystalline TiO_2 nanowires", Nano Letters, 2, 717–720, (2002).

[148] Y. Lin, "Photocatalytic activity of TiO_2 nanowire arrays", Materials Letters, 62, 1246–1248, (2008).

[149] Y. Lin, G.S. Wu, X.Y. Yuan, T. Xie, L.D. Zhang, "Fabrication and optical properties of TiO_2 nanowire arrays made by sol–gel electrophoresis deposition into anodic alumina membranes", Journal of Physics: Condensed Matter, 15, 2917–2922, (2003).

[150] L. Francioso, P. Siciliano, "Top–down contact lithography fabrication of a TiO_2 nanowire array over a SiO_2 mesa", Nanotechnology, 17, 3761–3767, (2006).

[151] K. Shantha Shankar, A.K. Raychaudhuri, "Fabrication of nanowires of multicomponent oxides: Review of recent advances", Materials Science and Engineering: C, 25, 738–751, (2005).

[152] T. Hyeon, J. Joo, S. Gu Kwon, T. Yu, M. Cho, J. Lee, J. Yoon, "Large–Scale Synthesis of TiO_2 Nanorods via Nonhydrolytic Sol–Gel Ester Elimination Reaction and Their Application to Photocatalytic Inactivation of E. coli", Journal of Physical Chemistry B, 109, 15297–15302, (2005).

[153] K. Hashimoto, Q. Wei, K. Hirota, K. Tajima, "Design and synthesis of TiO_2 nanorod assemblies and their application for photovoltaic devices", Chemistry of Materials, 18 (21), 5080–5087, (2006).

[154] L. Miao, S. Tanemura, S. Toh, K. Kaneko, M. Tanemura, "Fabrication, characterization and Raman study of anatase–TiO_2 nanorods by a heating–sol–gel template process", Journal of Crystal Growth, 264, 246–252, (2004).

[155] B. Koo, J. Park, Y. Kim, S.H. Choi, Y. Sung, T. Hyeon, "Simultaneous phase– and size–controlled synthesis of TiO_2 nanorods via non–hydrolytic Sol–Gel reaction of syringe pump delivered precursors", Journal of Physical Chemistry B, 110 (48), 24318–24323, (2006).

[156] L. Miao, S. Tanemura, S. Toh, K. Kaneko, M. Tanemura, "Heating–sol–gel template process for the growth of TiO_2 nanorods with rutile and anatase structure", Applied Surface Science, 238, 175–179, (2004).

[157] B.I. Lee, R.C. Bhave, "Experimental variables in the synthesis of brookite phase TiO_2 nanoparticles", Materials Science and Engineering A, 467, 146–149, (2007).

[158] M. Koelsch, S. Cassaignon, J.F. Guillemoles, J.P. Jolivet, "Comparison of optical and electrochemical properties of anatase and brookite TiO_2 synthesized by the sol–gel method", Thin Solid Films, 403–404, 312–319, (2002).

[159] S.R. Hall, V.M. Swinerd, F.N. Newby, A.M. Collins, S. Mann, "Fabrication of porous titania (Brookite) microparticles with complex morphology by Sol–Gel replication of pollen grains", Chemistry of Materials, 18(3), 598–600, (2006).

[160] Y. Djaoued, R. Brüning, D. Bersani, P.P. Lottici, S. Badilescu, "Sol–gel nanocrystalline brookite–rich titania films", Materials Letters, 58, 2618–2622, (2004).

[161] S.L. Isley, R.L. Penn, "Relative brookite and anatase content in Sol–Gel–synthesized titanium dioxide nanoparticles", Journal of Physical Chemistry B, 110(31), 15134–15139, (2006).

[162] L. Gao, Q. Zhang, "Effects of amorphous contents and particle size on the photocatalytic properties of TiO_2 nanoparticles", Scripta Materialia, 44, 1195–1198, (2001).

[163] E. Beyers, P. Cool, E.F. Vansant, "Anatase Formation during the synthesis of mesoporous titania and its photocatalytic effect", Physical Chemistry B, 109 (20), 10081–10086, (2005).

[164] Y. Hu, H.L. Tsai, C.L. Huang, "Phase transformation of precipitated TiO_2 nanoparticles", Materials Science and Engineering A, 344, 209–214, (2003).

[165] T. Sugimoto, X. Zhou, A. Muramatsu, "Synthesis of uniform anatase TiO_2 nanoparticles by gel-sol method 4. Shape control", Journal of Colloid and Interface Science, 259, 53–61, (2003).

[166] T. Sugimoto, X. Zhou, A. Muramatsu, "Synthesis of uniform anatase TiO_2 nanoparticles by Gel–Sol method: 1. solution chemistry of $Ti(OH)_n^{(4-n)+}$ complexes", Journal of Colloid and Interface Science, 252, 339–346, (2002).

[167] T. Sugimoto, X. Zhou, "Synthesis of uniform anatase TiO_2 nanoparticles by the Gel–Sol method: 2. adsorption of OH^- ions to $Ti(OH)_4$ gel and TiO_2 particles", Journal of Colloid and Interface Science, 252, 347–353, (2002).

[168] T. Sugimoto, "Underlying mechanisms in size control of uniform nanoparticles", Journal of Colloid and Interface Science, 309, 106–118, (2007).

[169] T. Sugimoto, X. Zhou, A. Muramatsu, "Synthesis of uniform anatase TiO_2 nanoparticles by gel–sol method 3. Formation process and size control", Journal of Colloid and Interface Science, 259, 43–52, (2003).

[170] J.M. Macak, F. Schmidt–Stein, P. Schmuki, "Efficient oxygen reduction on layers of ordered TiO_2 nanotubes loaded with Au nanoparticles", Electrochemistry Communications, 9, 1783–1787, (2007).

[171] Z. Chen, Y. Tang, H. Yang, Y. Xia, F. Li, T. Yi, C. Huang, "Nanocrystalline TiO_2 film with textural channels: Exhibiting enhanced performance in quasi-solid/solid–state dye–sensitized solar cells", Journal of Power Sources, 171, 990–998, (2007).

[172] M. Grätzel, J.J. Lagref, M.K. Nazeeruddin, "Artificial photosynthesis based on dye–sensitized nanocrystalline TiO_2 solar cells", Inorganica Chimica Acta, 361, 735–745, (2007).

[173] H. Kusama, M. Kurashige, K. Sayama, M. Yanagida, H. Sugihara, "Improved performance of Black–dye–sensitized solar cells with nanocrystalline anatase TiO_2 photoelectrodes prepared from $TiCl_4$ and ammonium carbonate", Journal of Photochemistry and Photobiology A: Chemistry, 189, 100–104, (2007).

[174] J. Wu, P. Li, S. Hao, H. Yang, Z. Lan, "A polyblend electrolyte $(PVP/PEG+KI+I_2)$ for dye–sensitized nanocrystalline TiO_2 solar cells", Electrochimica Acta, 52, 5334–5338, (2007).

[175] P.T. Hsiao, K. Wang, C. Cheng, H. Teng, "Nanocrystalline anatase TiO_2 derived from a titanate–directed route for dye–sensitized solar cells", Journal of Photochemistry and Photobiology A: Chemistry, 188, 19–24, (2007).

[176] R. Katoh, A. Furube, M. Murai, Y. Tamaki, K. Hara, M. Tachiya, "Effect of excitation wavelength on electron injection efficiency in dye–sensitized nanocrystalline TiO_2 and ZrO_2 films", Comptes Rendus Chimie, 9, 639–644, (2006).

[177] A. Furube, R. Katoh, T. Yoshihara, K. Hara, S. Murata, H. Arakawa, M. Tachiya, "Stepwise electron injection in the dye–sensitized nanocrystalline films of ZnO and TiO_2 with novel coumarin dye", Femtochemistry and Femtobiology, 525–528, (2004).

[178] W. Tai, "Photoelectrochemical properties of ruthenium dye–sensitized nanocrystalline SnO_2:TiO_2 solar cells", Solar Energy Materials and Solar Cells, 76, 65–73, (2003).

[179] J. Xia, N. Masaki, K. Jiang, S. Yanagida, "Fabrication and characterization of thin Nb_2O_5 blocking layers for ionic liquid–based dye–sensitized solar cells", Journal of Photochemistry and Photobiology A: Chemistry, 188, 120–127, (2007).

[180] A. Turkovi, Z. Cmjak Orel, "Dye–sensitized solar cell with CeO_2 and mixed CeO_2/SnO_2 photoanodes", Solar Energy Materials and Solar Cells, 45, 275–281, (1997).

[181] J. Bandara, H.C. Weerasinghe, "Enhancement of photovoltage of dye–sensitized solid–state solar cells by introducing high–band–gap oxide layers", Solar Energy Materials and Solar Cells, 88, 341–350, (2005).

[182] K.M.P. Bandaranayake, M.K. Indika Senevirathna, P.M.G.M. Prasad Weligamuwa, K. Tennakone, "Dye–sensitized solar cells made from nanocrystalline TiO_2 films coated with outer layers of different oxide materials", Coordination Chemistry Reviews, 248, 1277–1281, (2004).

[183] Y. Saito, S. Kambe, T. Kitamura, Y. Wada, Shozo Yanagida "Morphology control of mesoporous TiO_2 nanocrystalline films for performance of dye–sensitized solar cells", Solar Energy Materials and Solar Cells, 83, 1–13, (2004).

[184] S. Yanagida, G.K.R. Senadeera, K. Nakamura, T. Kitamura and Y. Wada, "Polythiophene–sensitized TiO_2 solar cells", Journal of Photochemistry and Photobiology A: Chemistry, 166, 75–80, (2004).

[185] Y. Saito, W. Kubo, T. Kitamura, Y. Wada, S. Yanagida, "I^-/I_3^- redox reaction behavior on poly(3, 4–ethylenedioxythiophene) counter electrode in dye–

sensitized solar cells", Journal of Photochemistry and Photobiology A: Chemistry, 164, 153–157, (2004).

[186] Y. Saito, T. Kitamura, Y. Wada, S. Yanagida, "Application of Poly(3,4–ethylenedioxythiophene) to counter electrode in dye–sensitized solar cells", Chemistry Letters, 31, 1060 (2002).

[187] M. Grätzel, "Conversion of sunlight to electric power by nanocrystalline dye–sensitized solar cells", Journal of Photochemistry and Photobiology A: Chemistry, 164, 3–14, (2004).

[188] M. Grätzel, "Photovoltaic performance and long–term stability of dye–sensitized meosocopic solar cells", Comptes Rendus Chimie, 9, 578–583, (2006).

[189] C. Bauer, G. Boschloo, E. Mukhtar, A. Hagfeldt, "Interfacial electron–transfer dynamics in ru(tcterpy)(NCS)$_3$–sensitized TiO$_2$ nanocrystalline solar cells", Journal of Physical Chemistry B, 106, 12693–12704, (2002).

[190] F. Pichot, J.R. Pitts, B.A. Gregg, "Low–Temperature sintering of TiO$_2$ colloids: application to flexible dye–sensitized solar cells", Langmuir, 16(13), 5626–5630, (2000).

[191] A.J. Frank, N. Kopidakis, J. van de Lagemaat, "Electrons in nanostructured TiO$_2$ solar cells: transport, recombination and photovoltaic properties (review)", Coordination Chemistry Reviews, 248, 1165–1179, (2004).

[192] S. Gunes, H. Neugebauer, N.S. Sariciftci, "Conjugated polymer–based organic solar cells", Chemical Reviews, 107(4), 1324–1338, (2007).

[193] G.K. Mor, K. Shankar, M. Paulose, O.K. Varghese, C.A. Grimes, "Use of highly-ordered TiO$_2$ nanotube arrays in dye–sensitized solar cells", Nano Letter, 6(2), 215–218, (2006).

[194] R. Katoh , A. Furube , A.V. Barzykin , H. Arakawa , M. Tachiya, "Kinetics and mechanism of electron injection and charge recombination in dye–sensitized nanocrystalline semiconductors (review)", Coordination Chemistry Reviews 248, 1195–1213, (2004).

[195] J. He, R. Mosurkal, L. Samuelson, L. Li, J. Kumar, "Dye–sensitized solar cell fabricated by electrostatic layer–by–layer assembly of amphoteric TiO$_2$ nanoparticles", Langmuir, 19(6), 2169–2174, (2003).

[196] J. Kallioinen , H. Santa–Nokki, T. Kololuoma , V. Tuboltsev, J. Korppi-Tommola, "Dynamic preparation of TiO$_2$ films for fabrication of dye–sensitized solar cells", Journal of Photochemistry and Photobiology A: Chemistry, 182, 187–191, (2006).

[197] J. Choy, S. Paek , H. Jung , Y. Lee , N. Park, S. Hwang, "Nanostructured TiO$_2$ films for dye–sensitized solar cells", Journal of Physic and Chemistry of Solids 67,1308–1311, (2006).

[198] M. Lim, S. Jang, R. Vittal, J. Lee, K. Kim, "Linking of N3 dye with C60 through diaminohydrocarbons for enhanced performance of dye–sensitized TiO$_2$ solar cells", Journal of Photochemistry and Photobiology A: Chemistry, 190, 128–134, (2007).

[199] J. Jiu, S. Isoda, M. Adachi, F. Wang, "Preparation of TiO$_2$ nanocrystalline with 3–5 nm and applicationfor dye–sensitized solar cell", Journal of Photochemistry and Photobiology A: Chemistry, 189, 314–321, (2007).

[200] C. Grimes, M. Paulose, K. Shankar, O. Vrghese, G. Mor, "Application of highly–ordered TiO$_2$ nanotube–arrays in heterojunction dye–sensitized solar cells" Journal of Physics D: Applied Physics, 39, 2498–2503, (2006).

[201] C. Grimes, "Synthesis and application of highly ordered arrays of TiO$_2$ nanotubes", Journal of Materials Chemistry, 17, 1451–1457, (2007).

[202] M. Anpo, M. Matsuoka, M. Kitano, M. Takeuchi, K. Tsujimaru, J. Thomas, "Photocatalysis for new energy production recent advances in photocatalytic water splitting reactions for hydrogen production",Catalysis Today, 122, 51–61, (2007).

[203] M. Ni, M. Leung, D. Leung, K. Sumathy, "A review and recent developments in photocatalytic water–splitting using TiO$_2$ for hydrogen production", Renewable and Sustainable Energy Reviews, 11, 401–425, (2007).

[204] H. Imahori, Y. Mori, Y. Matano, "Nanostructured artificial photosynthesis", Journal of Photochemistry and Photobiology C: Photochemistry Reviews, 4, 51–83, (2003).

[205] S. Takabayashi, R. Nakamura, Y. Nakato, "Nano–modified Si/TiO$_2$ composite electrode for efficient solar water splitting", Journal of Photochemistry and Photobiology A: Chemistry, 166, 107–113, (2004).

[206] A. Fujishima, K. Honda, "Electrochemical photolysis of water at a semiconductor electrode", Nature, 238, 37–38, (1972).

[207] P.R. Mishra, P.K. Shukla, O.N. Srivastava, "Study of modular PEC solar cells for photoelectrochemical splitting of water employing nanostructured TiO$_2$ photoelectrodes", International Journal of Hydrogen Energy, 32, 1680–1685, (2007).

[208] Mridula Misra, R.N. Pandey, O.N. Srivastava, "Solar hydrogen production employing n– TiO$_2$/Ti SC–SEP, photoelectrochemical solar cell", International Journal of Hydrogen Energy, 22, 501–508, (1997).

[209] S.K. Poznyak, A.I. Kokorin, A.I. Kulak, "Effect of electron and hole acceptors on the photoelectrochemical behaviour of nanocrystalline microporous TiO$_2$ electrodes", Journal of Electroanalytical Chemistry, 442, 99–105, (1998).

[210] S. Ichikawa, "Photoelectrocatalytic Production of Hydrogen from natural seawater under sunlight", Journal of Hydrogen Energy, 22, 615–618, (1997).

[211] J. Sung Lee, S. Ji, H. Jun, J. Jang, H. Son, P. Borse, "Photocatalytic hydrogen production from natural seawater", Journal of Photochemistry and Photobiology A: Chemistry, 189,141–144, (2007).

[212] A. Yamakata, T. Ishibashi, H. Onishi, "Kinetics of the photocatalytic water–splitting reaction on TiO$_2$ and Pt/TiO$_2$ studied by time–resolved infrared absorption spectroscopy", Journal of Molecular Catalysis A: Chemical, 199, 85–94, (2003).

[213] J. Nowotny, T. Bak, M.K. Nowotny, L.R. Sheppard, "TiO$_2$ surface active sites for water splitting", Journal of Physical Chemistry B, 110(37), 18492–18495, (2006).

[214] A. Galinska, J. Walendziewski, "Photocatalytic water splitting over Pt–TiO$_2$ in the presence of sacrificial reagents", Energy & Fuels, 19(3), 1143–1147, (2005).

[215] M. Ashokkumar, "An overview on semiconductor particulate systems for photoproduction of hydrogen", International Journal of Hydrogen Energy, 23, 427–438, (1998).

[216] K. Sayama, H. Arakawa, "Effect of Na$_2$CO$_3$ addition on photocatalytic decomposition of liquid water over various semiconductors catalysis", Journal of Photochemistry and Photobiology A: Chemistry, 77, 243–247, (1994).

[217] Li. FB, Li. XZ, "The enhancement of photodegradation efficiency using Pt–TiO$_2$ catalyst", Chemosphere, 48, 1103–1111, (2002).

[218] S. Jin, F. Shiraishi, "Photocatalytic activities enhanced for decompositions of organic compounds over metal photodepositing titanium dioxide", Chemical Engineering Journal, 97, 203–211, (2004).

[219] A.J. Bard, J.H. Park, S. Kim, "Novel carbon–doped TiO$_2$ nanotube arrays with high aspect ratios for efficient solar water splitting", Nano Letters, 6(1), 24–28, (2006).

[220] M. Misra, S.K. Mohapatra, V.K. Mahajan, K.S. Raja, "Design of a highly efficient photoelectrolytic cell for hydrogen generation by water splitting: application of TiO$_{2-x}$C$_x$ nanotubes as a photoanode and Pt/TiO$_2$ nanotubes as a cathode", Journal of Physical Chemistry C, 111(24), 8677–8685, (2007).

[221] M. Anpo, M. Kitano, M. Takeuchi, M. Matsuoka, J.M. Thomas, "Preparation of visible light–responsive TiO$_2$ thin film photocatalysts by an RF magnetron sputtering deposition method and their photocatalytic reactivity", Chemistry Letters, 34, 616–618, (2005).

[222] M. Anpo, M. Kitano, M. Matsuoka, M. Ueshima, "Recent developments in titanium oxide–based photocatalysts", Applied Catalysis A: General, 325, 1–14, (2007).

[223] M. Anpo, M. Kitano, K. Tsujimaru, "Decomposition of water in the separate evolution of hydrogen and oxygen using visible light–responsive TiO$_2$ thin film photocatalysts: Effect of the work function of the substrates on the yield of the reaction"Applied Catalysis A: General, 314, 179–183, (2006).

[224] P. Corbo, F. Migliardini, O. Veneri "Experimental analysis and management issues of a hydrogen fuel cell system for stationary and mobile application", Energy Conversion and Management, 48, 2365–2374, (2007).

[225] F. Walsh, D. Bavykin, A. Lapkin, P. Plucinski, J. Friedrich, "Reversible storage of molecular hydrogen by sorption into multilayered TiO$_2$ nanotubes", Journal of Physical Chemistry B, 109, 19422–19427, (2005).

[226] E. David, "An overview of advanced materials for hydrogen storage", Journal of Materials Processing Technology, 162–163, 169–177, (2005).

[227] B. Sakintuna, F. Lamari–Darkrim, M. Hirscher, "Metal hydride materials for solid hydrogen storage: A review", International Journal of Hydrogen Energy, 32, 1121–1140, (2007).

[228] D. Kim, Y. Park, H. Lee, "Tuning clathrate hydrates: Application to hydrogen storage", Catalysis Today, 120, 257–261, (2007).

[229] W.L. Mao, H. Mao, "Hydrogen storage in molecular compounds", Proceeding of the National Academy of Sciences of the United States of America, 101, 708–710, (2004).

[230] T.M. Inerbaev, V.R. Belosludov, R.V. Belosludov, M. Sluiter, Y. Kawazoe "Dynamics and equation of state of hydrogen clathrate hydrate as a function of cage occupation", Computational Materials Science, 36, 229–233, (2006).

[231] T.A. Strobel, C.J. Taylor, K.C. Hester, S.F. Dec, C.A. Koh, K.T. Miller, E.D. Sloan, "Molecular hydrogen storage in binary THF–H_2 clathrate hydrates", Journal of Physical Chemistry B., 110(34), 17121–17125, (2006).

[232] T. Kasuga, M. Hiramatsu, A. Hoson, T. Sekino, K. Niihara, "Titania nanotubes prepared by chemical processing", Advances Materials, 11, 1307–1311, (1999).

[233] X. Sun, Y. Li, "Synthesis and characterization of ion–exchangeable titanate nanotubes", Chemistry European Journal, 9, 2229 – 2238, (2003).

[234] D.V. Bavykin, V.N. Parmon, A.A. Lapkin, F.C. Walsh, "The effect of hydrothermal conditions on the mesoporous structure of TiO_2 nanotubes" Journal of Materials chemistry, 14, 3370–3377, (2004).

[235] V. Idakiev, Z.–Y. Yuan, T. Tabakova, B.–L. Su, "Titanium oxide nanotubes as supports of nano–sized gold catalysts for low temperature water–gas shift reaction", Applied Catalysis A: General, 281, 149–155, (2005).

[236] X. Chen, S. Mao, "Synthesis of titanium dioxide (TiO_2) nanomaterials", Journal of nanoscience and nanotechnology, 6 (4), 906–925, (2006).

[237] Y. Jia, A. Kleinhammes, H. Kulkarni, K. McGuire, L. McNeil, Y. Wu, "Synthesis and characterization of TiO_2 nanotube/hydroquinone hybrid structure", Journal of nanoscience and nanotechnology, 7 (2), 458–462, (2007).

[238] S.M. Yang, M.D. Lu, "Syntheses of titania and poly(3, 4–ethylenedioxythiophene) bilayer nanotubes", Journal of nanoscience and nanotechnology, 6 (12), 3960–3964, (2006).

[239] D. Guin, S.V. Manorama, J. Latha, S. Singh, "Photoreduction of silver on bare and colloidal TiO_2 nanoparticles/nanotubes: synthesis, characterization and tested for antibacterial outcome" Journal of Physical Chemistry C, 111, 13393–13397, (2007).

[240] J.N. Nian, H. Teng, "Hydrothermal synthesis of single–crystalline anatase TiO_2 nanorods with nanotubes as the precursor", Journal of Physical Chemistry B, 110(9), 4193–4198, (2006).

[241] Y. Chen, J.C. Crittenden, S. Hackney, L. Sutter, D. W. Hand, "Preparation of a novel TiO_2–based p–n junction nanotube photocatalyst", Environmental Science and Technology, 39(5), 1201–1208, (2005).

[242] S.H.S. Zein, A.R. Mohamed, "Mn/Ni/TiO$_2$ catalyst for the production of hydrogen and carbon nanotubes from methane decomposition", Energy & Fuels, 18(5), 1336–1345, (2004).

[243] M. Paulose, K. Shankar, S. Yoriya, H.E. Prakasam, O. Varghese, G.K. Mor, T.A. Latempa, A. Fitzgerald, C.A. Grimes, "Anodic growth of highly ordered TiO$_2$ nanotube arrays to 134 μm in length", Journal of Physical Chemistry B, 110(33), 16179–16184, (2006).

[244] C. Ruan, M. Paulose, O.K. Varghese, G.K. Mor, C.A. Grimes, "Fabrication of highly ordered TiO$_2$ nanotube arrays using an organic electrolyte", Journal of Physical Chemistry B, 109(33), 15754–15759, (2005).

[245] Z. Zhang, Y. Yuan, G. Shi, Y. Fang, L. Liang, H. Ding, L. Jin, "Photoelectrocatalytic activity of highly ordered TiO$_2$ nanotube arrays electrode for azo dye degradation", Environmental Sience and Technology, 41(17), 6259–6263, (2007).

[246] P. Pillaia, K.S. Rajaa, M. Misra, "Electrochemical storage of hydrogen in nanotubular TiO$_2$ arrays", Journal of Power Sources, 161, 524–530, (2006).

[247] J. Lin, S. Lim, J. Luo, Z. Zhong, W. Ji, "Room–temperature hydrogen uptake by TiO$_2$ nanotubes", Inorganic Chemistry, 44 (12), 4124 –4126, (2005).

[248] D.M. Antonelli, X.Hu, B.O. Skadtchenko, M. Trudeau, "Hydrogen storage in chemically reducible mesoporous and microporous Ti oxides", Journal of American Chemical Society, 128(36); 11740–11741, (2006).

[249] E. Balaur, J.M. Macak, H. Tsuchiya, P. Schmuki, "Wetting behaviour of layers of TiO$_2$ nanotubes with different diameters", Journal of Materials Chemistry, (42), 4488–4491, (2005).

[250] G.T. Wang, J.P. Tu, W.K. Zhang, X.L. Wang, H. Huang, X. P. Gan, "Photoassisted charge behavior of hydrogen storage alloy–TiO$_2$ /Pt electrodes", Journal of Physical Chemistry B, 109(27), 13210–13213, (2005).

[251] V.N.T. Kuchibhatla, A.S. Karakoti, D. Bera, S. Seal, "One dimensional nanostructured materials", Progress in Materials Science,52, 699–913, (2007).

[252] K.S. Johnson, J.A. Needoba, S.C. Riser, W.J. Showers, "Chemical sensor networks for the aquatic environment", Chemical Reviews, 107(2), 623–640, (2007).

[253] Q. Kuang, C. Lao, Z. L.Wang, Z. Xie, L. Zheng, "High–Sensitivity humidity sensor based on a single SnO$_2$ nanowire", Journal of American Chemical Society, 129(19), 6070–6071, (2007).

[254] J.T. McCue, J.Y. Ying, "SnO$_2$–In$_2$O$_3$ nanocomposites as semiconductor gas sensors for CO and NO$_x$ detection", Chemistry of Materials, 19(5), 1009–1015, (2007).

[255] X.D. Wang, J. Zhou, J.H. Song, J. Liu, N. Xu, Z.L. Wang, "Piezoelectric field effect transistor and nanoforce sensor based on a single ZnO nanowire", Nano Letters, 6(12), 2768–2772, (2006).

[256] F. Menil, V. Coillard, C. Lucat, "Critical review of nitrogen monoxide sensors for exhaust gases of lean burn engines", Sensors and Actuators B, 67, 1–23, (2000).

[257] L. Francioso, D.S. Presicce, M. Epifani, P. Siciliano, A. Ficarella, "Response evaluation of TiO_2 sensor to flue gas on spark ignition engine and in controlled environment", Sensors and Actuators B, 107, 563–571, (2005).

[258] S. Liu, A. Chen, "Coadsorption of horseradish peroxidase with thionine on TiO_2 nanotubes for biosensing", Langmuir, 21 (18), 8409 –8413, (2005).

[259] I. Fernández, A. Cremades, J. Piqueras, "Cathodoluminescence study of defects in deformed (110) and (100) surfaces of TiO_2 single crystals", Semiconductor Science and Technology, 20, 239–243, (2005).

[260] B. Ding, J. Kim, E. Kimura S. Shiratori, "Layer–by–layer structured films of TiO_2 nanoparticles and poly(acrylic acid) on electrospun nanofibres", Nanotechnology, 15, 913–917, (2004).

[261] R. Könenkamp, R. Word, M. Godinez, "Electroluminescence in nanoporous TiO_2 solid–state heterojunctions", Nanotechnology, 17 1858–1861, (2006).

[262] F. Haghighat, A. Khodadadi, Y. Mortazavi, "Temperature–independent ceria– and Pt–doped nano–size TiO_2 oxygen lambda sensor using Pt/SiO_2 catalytic filter", Sensors and Actuators B: Chemical, 129, 47–52, (2008).

[262] M.R. Mohammadi, D.J. Fray, M.C. Cordero–Cabrera, "Sensor performance of nanostructured TiO_2 thin films derived from particulate sol–gel route and polymeric fugitive agents", Sensors and Actuators B: Chemical, 124, 74–83, (2007).

[263] M.C. Carotta, M. Ferroni, D. Gnani, V. Guidi, M. Merli, G. Martinelli, M.C. Casale, M. Notaro, "Nanostructured pure and Nb–doped TiO_2 as thick film gas sensors for environmental monitoring", Sensors and Actuators B: Chemical, 58, 310–317, (1999).

[264] F. Edelman, H. Hahn, S. Seifried, C. Alof, H. Hoche, A. Balogh, P. Werner, K. Zakrzewska, M. Radecka, P. Pasierb, "Structural evolution of SnO_2–TiO_2 nanocrystalline films for gas sensors", Materials Science and Engineering B, 69–70, 386–391, (2000).

[265] O.K. Tan, W. Cao, W. Zhu, J.W. Chai, J.S. Pan, "Ethanol sensors based on nano–sized α–Fe_2O_3 with SnO_2, ZrO_2, TiO_2 solid solutions", Sensors and Actuators B: Chemical, 93, 396–401, (2003).

[265] A.M. Ruiz, A. Cornet, J. Morante, "Study of La and Cu influence on the growth inhibition and phase transformation of nano–TiO_2 used for gas sensors", Sensors and Actuators B: Chemical, 100, 256–260, (2004).

[267] S.H. Si, Y.S. Fung, D.R. Zhu, "Improvement of piezoelectric crystal sensor for the detection of organic vapors using nanocrystalline TiO_2 films", Sensors and Actuators B: Chemical, 108, 165–171, (2005).

[268] G.N. Chaudhari, A.M. Bende, A.B. Bodade, S.S. Patil, V.S. Sapkal, "Structural and gas sensing properties of nanocrystalline TiO_2:WO_3–based hydrogen sensors", Sensors and Actuators B: Chemical, 115, 297–302, (2006).

[269] R. Wu, Y. Sun, C. Lin, H. Chen, M. Chavali, "Composite of TiO_2 nanowires and Nafion as humidity sensor material", Sensors and Actuators B: Chemical, 115, 198–204, (2006).

[270] P. Su, Y. Sun, C. Lin, "Novel low humidity sensor made of TiO_2 nanowires/poly(2–acrylamido–2–methylpropane sulfonate) composite material film combined with quartz crystal microbalance", Talanta, 69, 946–951, (2006).

[271] C. Han, D. Hong, S. Han, J. Gwak, K.C. Singh, "Catalytic combustion type hydrogen gas sensor using TiO_2 and UV–LED", Sensors and Actuators B: Chemical, 125, 224–228, (2007).

[272] Y. Wang, X. Du, Y. Mu , L. Gui, P. Wang,, Y. Tang, "A new highly selective H2 sensor based on TiO_2/PtO–Pt dual–layer films", Chemistry of Materials,14, 3953–3957, (2002).

[273] H. Miyazaki, T. Hyodo, Y. Shimizu, M. Egashira, "Hydrogen–sensing properties of anodically oxidized TiO_2 film sensors: Effects of preparation and pretreatment conditions"Sensors and Actuators B: Chemical, 108, 467–472, (2005).

[274] O.K. Varghese, D. Gong, M. Paulose, K.G. Ong, C.A. Grimes, "Hydrogen sensing using titania nanotubes", Sensors and Actuators B, 93, 338–344, (2003).

[275] A.M. Taurino, S. Capone, P. Siciliano, T. Toccoli, A. Boschetti, L. Guerini, S. Iannotta, "Nanostructured TiO_2 thin films prepared by supersonic beams and their application in a sensor array for the discrimination of VOC", Sensors and Actuators B, 92, 292–302, (2003).

[276] A. M. Taurino, R. Rella, J. Spadavecchia, M.G. Manera, S. Capone, M. Martino, A.P. Caricato, T. Tunno, "Acetone and ethanol solid–state gas sensors based on TiO_2 nanoparticles thin film deposited by matrix assisted pulsed laser evaporation", Sensors and Actuators B, 127, 426–431, (2007).

[277] E.K. Suh, B. Karunagaran, Periyayya Uthirakumar, S.J. Chung, S. Velumani, "TiO_2 thin film gas sensor for monitoring ammonia", Materials Characterization, 58, 680–684, (2007).

[278] Y. Shimizu, T. Okamoto, Y. Takao, M. Egashira, "Desorption behavior of ammonia from TiO_2–based specimens–ammonia sensing mechanism of double–layer sensors with TiO_2–based catalyst layers", Journal of Molecular Catalysis A: Chemical, 155, 183–191, (2000).

[279] M. Guarino, A. Costa, M. Porro, "Photocatalytic TiO_2 coating–to reduce ammonia and greenhouse gases concentration and emission from animal husbandries", Bioresource Technology, 99, 2650–2658, (2008).

[280] S. Lee, D. Yang, T. Kunitake, "Regioselective imprinting of anthracenecarboxylic acids onto TiO_2 gel ultrathin films: an approach to thin film sensor", Sensors and Actuators B: Chemical, 104, 35–42, (2005).

[281] D. Xiao, X. Shu, Y. Chen, H. Yuan, S. Gao, "H_2O_2 sensor based on the room–temperature phosphorescence of nano TiO_2/SiO_2 composite", Analytical Chemistry, 79, 3695–3702, (2007).

[282] G. Wang, Q. Wang, W. Lu, J. Li, "Photoelectrochemical study on charge transfer properties of TiO_2–B nanowires with an application as humidity sensors", Journal of Physical Chemistry B, 110, 22029–22034, (2006).

[283] K.P. Biju, M.K. Jain, "Effect of polyethylene glycol additive in sol on the humidity sensing properties of a TiO_2 thin film", Measurement Science & Technology, 18, 2991–2996, (2007).

[284] F. Al–Momani, E. Touraud, J.R. Degorce–Dumas, J. Roussy, O. Thomas, "Biodegradability enhancement of textile dyes and textile wastewater by VUV photolysis", Journal of Photochemistry and Photobiology A: Chemistry, 153, 191–197, (2002).

[285] J. Li, L. Zheng, L. Li, G. Shi, Y. Xian, L. Jin, "Photoelectro–synergistic catalysis combined with a FIA system application on determination of chemical oxygen demand", Talanta, 72, 1752–1756, (2007).

[286] J. Li, L. Zheng, L. Li, G. Shi, Y. Xian, L. Jin, "Determination of chemical oxygen demand using flow injection with Ti/ TiO_2 electrode prepared by laser anneal", Measurement Science & Technology, 18, 945–951, (2007).

[287] J. Chen, J. Zhang, Y. Xian, X. Ying, M. Liua, L. Jin, "Preparation and application of TiO_2 photocatalytic sensor for chemical oxygen demand determination in water research", Water Research, 39,1340–1346, (2005).

[288] L. Francioso, D.S. Presicce, P. Siciliano, A. Ficarella, "Combustion conditions discrimination properties of Pt–doped TiO_2 thin film oxygen sensor", Sensors and Actuators B: Chemical, 123, 516–521, (2007).

[289] B. Elyassi, N. Rajabbeigi, A.A. Khodadadi, S. Mohajerzadeh, Y. Mortazavi, M. Sahimi, "Oxygen sensor with solid–state CeO_2–ZrO_2–TiO_2 reference" Sensors and Actuators B: Chemical, 108, 341–345, (2005).

[290] L. Zheng, M. Xu, T. Xu, "TiO_{2-x} thin films as oxygen sensor", Sensors and Actuators B: Chemical, 66, 28–30, (2000).

[291] H. Lee, W. Hwang, "Substrate effects on the oxygen gas sensing properties of SnO_2/TiO_2 thin films", Applied Surface Science, 253, 1889–1897, (2006).

[292] K. Zakrzewska, "Gas sensing mechanism of TiO_2–based thin films", Vacuum, 74, 335–338, (2004).

[293] T. Jantson, T. Avarmaa, H. Mändar, T. Uustare, R. Jaaniso, "Nanocrystalline Cr_2O_3–TiO_2 thin films by pulsed laser deposition", Sensors and Actuators B: Chemical, 109, 24–31, (2005).

[294] S. Zhuiykov, W. Wlodarski, Y. Li, "Nanocrystalline V_2O_5–TiO_2 thin–films for oxygen sensing prepared by sol–gel process", Sensors and Actuators B: Chemical, 77, 484–490, (2001).

[295] R.K. Sharma, M.C. Bhatnagar, G.L. Sharma, "Mechanism of highly sensitive and fast response Cr doped TiO_2 oxygen gas sensor", Sensors and Actuators B: Chemical, 45, 209–215, (1997).

[296] R.K. Sharma, M.C. Bhatnagar "Improvement of the oxygen gas sensitivity in doped TiO$_2$ thick films", Sensors and Actuators B: Chemical, 56, 215–219, (1999).

[297] M.Z. Atashbar, H.T. Sun, B. Gong, W. Wlodarski, R. Lamb, "XPS study of Nb–doped oxygen sensing TiO$_2$ thin films prepared by sol–gel method", Thin Solid Films, 326, 238–244, (1998).

[298] O.K. Varghese, C.A. Grimes, "Metal oxide nanoarchitectures for environmental sensing", Journal of Nanoscience and Nanotechnology, 3, 277–293, (2003).

[299] B. Min, S. Choi, "SnO$_2$ thin film gas sensor fabricated by ion beam deposition", Sensors and Actuators B: Chemical, 98, 239–246, (2004).

[300] J. Sheng, N. Yoshida, J. Karasawa, T. Fukami, "Platinum doped titania film oxygen sensor integrated with temperature compensating thermistor", Sensors and Actuators B: Chemical, 41, 131–136, (1997).

[301] M. Li, Y. Chen, "An investigation of response time of TiO$_2$ thin–film oxygen sensors", Sensors and Actuators B: Chemical, 32, 83–85, (1996).

[302] Y. Xu, K. Yao, X. Zhou, Q. Cao, "Platinum–titania oxygen sensors and their sensing mechanisms", Sensors and Actuators B: Chemical, 14, 492–494, (1993).

[303] U. Kirner, K.D. Schierbaum, W. Göpel, B. Leibold, N. Nicoloso, W. Weppner, D. Fischer, W.F. Chu, "Low and high temperature TiO$_2$ oxygen sensors", Sensors and Actuators B: Chemical, 1, 103–107, (1990).

[304] M. Winter, R.J. Brodd, "What are batteries, fuel cells and supercapacitors?", Chemical Reviews, 104(10), 4245–4270, (2004).

[305] M.S. Whittingham, R.F. Savinell, T. Zawodzinski, "Introduction: batteries and fuel cells", Chemical Reviews, 104(10), 4243–4244, (2004).

[306] C.P. Grey, N. Dupre, "NMR studies of cathode materials for lithium–ion rechargeable batteries", Chemical Reviews, 104(10), 4493–4512, (2004).

[307] M.S. Whittingham, "Lithium batteries and cathode materials", Chemical Reviews, 104(10), 4271–4302, (2004).

[308] L. Benco, J. Barras, M. Atanasov, C.A. Daul, E. Deiss, "First–principles prediction of voltages of lithiated oxides for lithium–ion batteries",Solid State Ionics,112, 255–259, (1998).

[309] K. Xu, "Nonaqueous liquid electrolytes for lithium–based rechargeable batteries", Chemical Reviews, 104(10), 4303–4418, (2004).

[310] Y. Yang, Z. Zhang, Z. Gong, "Electrochemical performance and surface properties of bare and TiO$_2$–coated cathode materials in lithium–ion batteries", Journal of Physical chemistry B, 108(45), 17546–17552, (2004).

[311] C.W. Lin, C.L. Hung, M. Venkateswarlu, B.J. Hwang, "Influence of TiO$_2$ nano–particles on the transport properties of composite polymer electrolyte for lithium–ion batteries", Journal of Power Sources, 146, 397–401, (2005).

[312] J. Xu, C. Jia, B. Cao, W.F. Zhang, "Electrochemical properties of anatase TiO$_2$ nanotubes as an anode material for lithium–ion batteries", Electrochimica Acta, 52, 8044–8047, (2007).

[313] B. He, B. Dong, H. Li, "Preparation and electrochemical properties of Ag–modified TiO_2 nanotube anode material for lithium–ion battery", Electrochemistry Communications, 9, 425–430, (2007).

[314] Q. Wang, Z. Wen, J. Li, "Solvent–controlled synthesis and electrochemical lithium storage of one–dimensional TiO_2 nanostructures", Inorganic Chemistry, 45, 6944–6949, (2006).

[315] H. Huang, W.K. Zhang, X.P. Gan, C. Wang, L. Zhang, "Electrochemical investigation of TiO_2/carbon nanotubes nanocomposite as anode materials for lithium–ion batteries", Materials Letters, 61, 296–299, (2007).

[316] L.J. Fu, H. Liu, H.P. Zhang, C. Li, T. Zhang, Y.P. Wu, H.Q. Wu, "Novel TiO_2/C nanocomposites for anode materials of lithium ion batteries", Journal of Power Sources, 159, 219–222, (2006).

[317] V. Subramanian, A. Karki, K.I. Gnanasekar, Fannie Posey Eddy, B. Rambabu, "Nanocrystalline TiO_2 (anatase) for Li–ion batteries", Journal of Power Sources 159, 186–192, (2006).

[318] A. Armstrong, G. Armstrong, J. Canales, P.G. Bruce, "TiO_2–B nanowires as negative electrodes for rechargeable lithium batteries", Journal of Power Sources 146, 501–506, (2005).

[319] L. Kavan, M. Kalba, M. Zukalova, I. Exnar, V. Lorenzen, R. Nesper, M. Graetzel, "Lithium storage in nanostructured TiO_2 made by hydrothermal growth", Chemistry of Materials, 16 (3), 477 –485, (2004).

[320] J. Li, Z. Tang, Z. Zhang, "Layered hydrogen Titanate nanowires with novel lithium intercalation properties", Chemistry of Materials, 17(23), 5848–5855, (2005).

[321] J. Li, Z. Tang, Z. Zhang, "Controllable formation and electrochemical properties of one–dimensional nanostructured spinel $Li_4Ti_5O_{12}$", Electrochemistry Communications, 7, 894–899, (2005).

[322] X. Gao, H. Zhu, G. Pan, S. Ye, Y. Lan, F. Wu, D. Song, "Preparation and electrochemical characterization of anatase nanorods for lithium–inserting electrode material", Journal of Physical Chemistry B, 108(9), 2868–2872, (2004).

[323] Y. Zhu, H. Li, Y. Koltypin, Y. R. Hacohen, A. Gedanken, "Sonochemical synthesis of titania whiskers and nanotubes", Chemical Communications, 2616–2617, (2001).

[324] K. Gurunathan, D.P. Amalnerkar, D.C. Trivedi, "Synthesis and characterization of conducting polymer composite (PAn/ TiO_2) for cathode material in rechargeable battery", Materials Letters, 57, 1642–1648, (2003).

[325] Mansoor M. Amiji, "Nanotechnology for cancer therapy", CRC Press, Taylor & Francis Group, (2007).

[326] M. Kalbacova, J.M. Macak, F. Schmidt–Stein, C.T. Mierke, P. Schmuki, "TiO_2 nanotubes: photocatalyst for cancer cell killing", Physical Status Solidi–Rapid Research Letters, 2, 194–196, (2008).

[327] A. Fujishima, R. Cai, Y. Kubota, T. Shuin, H. Sakai, K. Hashimoto, "Induction of cytotoxicity by photoexcited TiO$_2$ particles", Cancer Research, 52, 2346–2348, (1992).

[328] J. Cheon, J. Seo, H. Chung, M. Kim, J. Lee, I. Choi, "Development of water–soluble single–crystalline TiO$_2$ nanoparticles for photocatalytic cancer–cell treatment", Small, 3, 850–853, (2007).

[329] J. Cheon, J. Seo, Y. Jun, S. Ko, "In situ one–pot synthesis of 1–dimensional transition metal oxide nanocrystals", Journal of Physical Chemistry B, 109, 5389–5391, (2005).

[330] G.E. Woloschak, T. Paunesku , T. Rajh, G. Wiederrecht, J. Maser, S. Vogt, N. Stojicevic, M. Protic, B. Lai, J. Oryhon, M. Thurnauer "Biology of TiO$_2$–oligonucleotide nanocomposites", Nature Materials 2, 343–346, (2003).

[331] G.E. Woloschak, T. Paunesku, S. Vogt, B. Lai, J. Maser, N. Kenneth, T. Thurn, C. Osipo, H. Liu, D. Legnini, Z. Wang, C. Lee, "Intracellular distribution of TiO$_2$–DNA oligonucleotide nanoconjugates directed to nucleolus and mitochondria indicates sequence specificity", Nano Letters, 7 (3), 596–601, (2007).

[332] H.E. Sussman, F. Writer, "Nanodevices hold promise for gene therapy", Drug Discovery Today, 8, 564–565, (2003).

[333] T. Lopez, E. Ortiz, P. Quintana, R.D. Gonz´alez, "A nanostructured titania bioceramic implantable device capable of drug delivery to the temporal lobe of the brain", Colloids and Surfaces A: Physicochemical Engineering Aspects, 300, 3–10, (2007).

[334] A. Peterson, T. Lopez, E. Ortiz Islas, R.D. Gonzalez, "Pore structures in an implantable sol–gel titania ceramic device used in controlled drug release applications: A modeling study", Applied Surface Science, 253, 5767–5771, (2007).

[335] T. López, P. Quintana, E. Ortiz–Islas, E. Vinogradova, J. Manjarrez, D.H. Aguilar, P. Castillo–Ocampo, C. Magaña , J.A. Azamar, "Characterization of sodium phenytoin co–gelled with titania for a controlled drug–release system", Materials Characterization, 58, 823–828, (2007).

[336] M. h Tuszynski, "Growth–factor gene therapy for neurodegenerative disorders", The Lancet Neurology, 1, 51–57, (2002).

[337] T. Lo´pez, J. Sotelo, J. Navarrete, J.A. Ascencio, "Synthesis of TiO$_2$ nanostructured reservoir with temozolomide: Structural evolution of the occluded drug", Optical Materials, 29, 88–94, (2006).

[338] G.A. Mansoori, P. Mohazzabi, P. McCormack, S. Jabbari, "Nanotechnology in cancer prevention, detection and treatment: bright future lies ahead", World Review of Science, Technology and Sustainable Development, 4, 226–257, (2007).

[339] M. Ann English, B. Shen, J.C. Scaiano, "Zeolite encapsulation decreases TiO$_2$–photosensitized ROS generation in cultured human skin fibroblasts", Photochemistry and photobiology, 82, 5–12, (2006).

[340] D. Dondi, A. Albini, N. Serpone, "Interactions between different solar UVB/UVA filters contained in commercial suncreams and consequent loss of UV protection", Photochemistry and Photobiology Science, 5, 835–843, (2006).

[341] C. S. Sander, H. Chang, S. Salzmann, C.S.L. Muller, S. Ekanayake–Mudiyanselage, P. Elsner, J.J. Thiele, "Photoaging is associated with protein oxidation in human skin in vivo", Journal of Investigative Dermatology, 118, 618–625, (2002).

[342] N. Serpone, D. Dondi, A. Albini, "Inorganic and organic UV filters: Their role and efficacy in sunscreens and suncare products", Inorganica Chimica Acta, 360, 794–802, (2007).

[343] T. Kallio, S. Alajoki, V. Pore, M Ritala, J. Laine, M. Leskelä, P. Stenius, "Antifouling properties of TiO_2: Photocatalytic decomposition and adhesion of fatty and rosin acids, sterols and lipophilic wood extractives", Colloids and Surfaces A: Physicochemical and Engineering Aspects, 291, 162–176, (2006).

[344] A. Fujishima, K. Hashimoto, T. Watanabe, "TiO_2 photocatalysis fundaments and applications", University of Tokyo Published by BKC, Inc., Chiyoda–ku, Tokyo, (1999).

[345] Q. Cheng, C. Li, V. Pavlinek, P. Saha, H. Wang, "Surface–modified antibacterial TiO_2/Ag+ nanoparticles: Preparation and properties", Applied Surface Science, 252, 4154–4160, (2006).

[346] C.C. Trapalis, P. Keivanidis, G. Kordas, M. Zaharescu, M. Crisan, A. Szatvanyi, M. Gartner, "TiO_2 (Fe^{+3}) nanostructured thin films with antibacterial properties", Thin Solid Films 433, 186–190, (2003).

[347] H.J. Zhang, D.Z. Wen, "Antibacterial properties of Sb– TiO_2 thin films by RF magnetron co–sputtering", Surface & Coatings Technology, 201, 5720–5723, (2007).

[348] A. Mills, A. Lepre, N. Elliott, S. Bhopal, I.P. Parkin, S.A. O'Neill, "Characterisation of the photocatalyst Pilkington ActivTM: a reference film photocatalyst?", Journal of Photochemistry and Photobiology A: Chemistry, 160, 213–224, (2003).

[349] P.Chin, D.F. Ollis, "Decolorization of organic dyes on Pilkington ActivTM photocatalytic glass", Catalysis Today, 123, 177–188, (2007).

[350] T. Yuranova, D. Laub, J. Kiwi, "Synthesis, activity and characterization of textiles showing self–cleaning activity under daylight irradiation", Catalysis Today, 122, 109–117, (2007).

[351] E. Herrero, K. Franaszczuk, A. Wieckowski, "Electrochemistry of methanol at low index crystal planes of platinum: an integrated voltammetric and chronoamperometric study" Journal of Physical Chemistry, 98, 5074–5083, (1994).

[352] M. Hepel, I. Dela, T. Hepel, J. Luo, C.J. Zhong, "Novel dynamic effects in electrocatalysis of methanol oxidation on supported nanoporous TiO_2 bimetallic nanocatalysts", Electrochimica Acta, 52, 5529–5547, (2007).

[353] J.M. Macak, P.J. Barczuk, H. Tsuchiya, M.Z. Nowakowska, A. Ghicov, M. Chojak, S. Bauer, S. Virtanen, P.J. Kulesza, P. Schmuki, "Self–organized nanotubular TiO_2 matrix as support for dispersed Pt/Ru nanoparticles: Enhancement of the electrocatalytic oxidation of ethanol", Electrochemistry Communications, 7, 1417–1422, (2005).

[354] D. Xia, Z. Wang, G. Chen, L. Zhang, "Studies on the electrocatalytic properties of PtRu/C–TiO_2 toward the oxidation of methanol", Journal of Alloys and Compounds, 450, 148–151, (2008).

[355] F. Marken, E.V. Milsom, J. Novak, M. Oyama, "Electrocatalytic oxidation of nitric oxide at TiO_2–Au nanocomposite film electrodes", Electrochemistry Communications, 9, 436–442, (2007).

[356] C. Feng, N. Sugiura, S. Shimada, T. maekawa, "Development of a high performance electrochemical wastewater treatment system", Journal of Hazardous Materials,103, 65–78, (2003).

[357] A.R. Khataee, A. Aleboyeh, H. Aleboyeh, "Crystallite phase–controlled preparation, characterisation and photocatalytic properties of titanium dioxide nanoparticles", Journal of Experimental Nanoscience, 4 (2), 121–137, (2009).

[358] J.W. Verhoeven, "Glossary of terms used in photochemistry", Pure & Applied Chemistry, 68 (12), 2223–2286, (1996).

[359] V. Parmon, A.V. Emeline, N. Serpone, "Glosary of terms in photocatalysis and radiocatalysis", International Journal of Photochemistry, 4, 91–131, (2002).

[16/360] C.P. Poole Jr, F.J. Owens, "Introduction to nanotechnology", John Wiley & Sons, Inc, Hoboken. New Jersey. (2003).

[361] E. Stathatos , D. Tsiourvas, P. Lianos, "Titanium dioxide films made from reverse micelles and their use for the photocatalytic degradation of adsorbed dyes", Colloids and Surfaces A: Physicochemical and Engineering Aspects, 149 (1–3): 49–56, (1999) .

[362] O.V. Makarova, T. Rajh, M.C. Thurnauer, A. Martin, P.A. Kemme, D. Cropek, "Surface modification of TiO_2 nanoparticles for photochemical reduction of nitrobenzene", Environmental Science and Technology, 34(22), 4797–4803, (2000).

[363] M. Andersson, L. Osterlund, S. Ljungstrom, A. Palmqvist, " Preparation of nanosize anatase and rutile TiO_2 by hydrothermal treatment of microemulsions and their activity for photocatalytic wet oxidation of phenol", Journal of Physical Chemistry B, 106 (41): 10674–10679, (2002).

[364] J. Chen, M. Liu, L. Zhang, J. Zhang, L. Jin, "Application of nano TiO_2 towards polluted water treatment combined with electro–photochemical method", Water Research, 37 (16): 3815–3820, (2003).

[365] L. Li, W. Zhu, P. Zhang, Z. Chen, W. Han, "Photocatalytic oxidation and ozonation of catechol over carbon–black–modified nano–TiO_2 thin films supported on Al sheet", Water Research, 37 (15): 3646–3651, (2003).

[366] W. Fu, H. Yang, L. Chang, H–Bala, M. Li, G. Zou, "Anatase TiO_2 nanolayer coating on strontium ferrite nanoparticles for magnetic photocatalyst ", Colloids and Surfaces A: Physicochemical Engineering Aspects, 289, 47–52, (2006).

[367] I. Ilisz, A. Dombi, K. Mogyorósi, I. Dékány, " Photocatalytic water treatment with different TiO_2 nanoparticles and hydrophilic/hydrophobic layer silicate adsorbents", Colloids and Surfaces A: Physicochemical Engineering Aspects, 230 (1–3), 89–97, (2004).

[368] H. Wang, J.J. Miao, J.M. Zhu, H.M. Ma, J.J. Zhu, H.Y. Chen, " Mesoporous spherical aggregates of anatase nanocrystals with wormhole–like framework structures: Their chemical fabrication, characterization and photocatalytic performance ", Langmuir, 20 (26), 11738–11747, (2004).

[369] T. Peng, D. Zhao, K.D.W. Shi, K. Hirao, "Synthesis of titanium dioxide nanoparticles with mesoporous anatase wall and high photocatalytic activity", Journal of Physical Chemistry B, 109 (11), 4947–4952, (2005).

[370] S.M. Paek, H. Jung, Y.J. Lee, M. Park, S.J. Hwang, J.H. Choy, "Exfoliation and reassembling route to mesoporous Titania nanohybrids", Chemistry of Materials, 18(5), 1134–1140, (2006).

[371] W.A. Adams, M.G. Bakker, T.I. Quickenden, "Photovoltaic properties of ordered mesoporous silica thin film electrodes encapsulating titanium dioxide particles ", Journal of Photochemistry and Photobiology A: Chemistry , 81 (2–3), 166–173, (2006) .

[372] M. Asilturk, F. Sayýlkan , S. Erdemoglu , M. Akarsu , H. Sayýlkan, M. Erdemoglu E. Arpac, " Characterization of the hydrothermally synthesized nano–TiO_2 crystallite and the photocatalytic degradation of Rhodamine B" Journal of Hazardous Materials B,129 (1–3), 164–170, (2006).

[373] J. Wang , T. Ma, Z. Zhang , X. Zhang, Y. Jiang , D. Dong, P. Zhang, Ying. Li, "Investigation on the sonocatalytic degradation of parathion in the presence of nanometer rutile titanium dioxide (TiO_2) catalyst", Journal of Hazardous Materials, 137, 972–980, (2006).

[374] Y. Li, X. Li, J. Li, J. Yin, " Photocatalytic degradation of methyl orange by TiO_2–coated Activated carbon and kinetic study", Water research, 40(6): 1119–1126, (2006).

[375] C.H. Wu, H.W. Chang, J.M. Chern, "Basic dye decomposition kinetics in a photocatalytic slurry reactor", Journal of Hazardous Materials, 137, 2006, 336–343, (2006).

[376] N. Daneshvar, A.R. Khataee, "Removal of azo dye C.I. acid red 14 from contaminated water using fenton, UV/H_2O_2, $UV/H_2O_2/Fe(II)$, $UV/H_2O_2/Fe(III)$ and $UV/H_2O_2/Fe(III)/$ oxalate processes: A comparative study ", Journal of Environmental Science and Health – Part A, 41 (3), 315–328, (2006).

[377] R.R. Bacsa, J. Kiwi, "Effect of rutile phase on the photocatalytic properties of nanocrystalline titania during the degradation of p–coumaric acid", Applied Catalysis B: Environmental, 16, 19–29 (1998).

[378] R.I. Bickley, T. Gonzalez–Carreno, J.S. Lees, L. Palmisano, R.J.D. Tilley, "A structural investigation of titanium dioxide photocatalysts", Journal of Solid State Chemistry, 92, 178–190, (1991).

[379] V. Grassian, "Environmental catalysis", CRC Press, Taylor & Francis Group, (2005).

[380] D.C. Hurum, A.G. Agrios, K.A. Gray, T. Rajh, M.C. Thurnauer, "Explaining the enhanced photocatalytic activity of degussa P25 mixed–phase TiO_2 using EPR", Journal of Physical Chemistry B, 107(19), 4545–4549, (2003).

[381] D.C. Hurum, K.A. Gray, T. Rajh, M.C. Thurnauer, "Recombination pathways in the degussa P25 formulation of TiO_2: surface versus lattice mechanisms", Journal of Physical Chemistry B, 109(2), 977–980, (2005).

[382] D. Beydoun, R. Amal, G. Low, S. McEvoy, "Occurrence and prevention of photodissolution at the phase junction of magnetite and titanium oxide", Journal of Molecular Catalysis A: Chemical, 180: 193–200, (2002).

[383]. P.A. Sant, P.V. Kamat, "Inter–Particle electron transfer between size–quantized CdS and TiO_2 semiconductor nanoclusters", Physical Chemistry Chemical Physics, 4, 198–203, (2002).

[384]. J.J. Sene, W.A. Zeltner, M.A. Anderson, "Fundamental photoelectrocatalytic and electrophoretic mobility studies of TiO_2 and V–Doped TiO_2 thin–film electrode materials", Journal of Physical Chemistry B, 107, 1597–1603, (2003).

[385] Z. Ding, H.Y. Zhu, G.Q. Lu, P.F. Greenfield, "Photocatalytic properties of titania pillared clays by different drying methods", Journal of Colloid and Interface Science, 209 (1), 193–199, (1999).

[386] C. Ooka, H. Yoshida, K. Suzuki, T. Hattori, "Highly hydrophobic TiO_2 pillared clay for photocatalytic degradation of organic compounds in water", Microporous and Mesoporous Materials, 67, (2–3), 143–150, (2004).

[387] J.H. Choy, J.H. Park, J.H. Yang, J.B. Yoon, S.J. Hwang, "Intracrystalline structure and physicochemical properties of mixed SiO_2–TiO_2 sol–pillared aluminosilicate", Journal of Physical Chemistry B, 110(4), 1592–1598, (2006).

[388] H.Y. Zhu, J.Y. Li, J.C. Zhaob, G.J. Churchman, "Photocatalysts prepared from layered clays and titanium hydrate for degradation of organic pollutants in water", Applied Clay Science, 28 (1–4), 79–88, (2005).

[389] J. Feng, X. Hu, P.L. Yue, H.Y. Zhu, G.Q. Lu, "Degradation of azo–dye orange II by a photoassisted Fenton reaction using a novel composite of iron oxide and silicate nanoparticles as a catalyst", Industrial & Engineering Chemistry Research, 42(10), 2058–2066, (2003).

[390] R. Say, E. Birlik, A. Denizli, A. Ersöz, "Removal of heavy metal ions by dithiocarbamate–anchored polymer/organosmectite composites", Applied Clay Science, 31(3–4), 298–305, (2006).

[391] J.H. Choi, S.D. Kim, S.H. Noh, S.J. Oh, W.J. Kim, "Adsorption behaviors of nano–sized ETS–10 andAl–substituted–ETAS–10 in removing heavy metal ions, Pb^{2+} and Cd^{2+}", Microporous and Mesoporous Materials, 87 (3), 163–169, (2006).

[392] H.Y., Zhu, J.A. Orthman, J.Y. Li, J.C. Zhao, G.J. Churchman, E.F. Vansant, "Novel composites of TiO_2 (Anatase) and silicate nanoparticles", Chemistry of Materials., 14(12), 5037–5044, (2002).

[393] V. Belessi, D. Lambropoulou, I. Konstantinou, A. Katsoulidis, P. Pomonis, D. Petridis, T. Albanis, "Structure and photocatalytic performance of TiO_2/clay nanocomposites for the degradation of dimethachlor", Applied Catalysis B: Environmental, 73, 292–299, (2007).

[394] J. Feng, R.S.K. Wong, X. Hu, P. Yue, "Discoloration and mineralization of Orange II by using Fe^{3+}–doped TiO_2 and bentonite clay–based Fe nanocatalysts", Catalysis Today, 98, 441–446, (2004).

[395] K. Mogyorosi, A. Farkas, I. Dekany, I. Ilisz, A. Dombi, "TiO_2–Based photocatalytic degradation of 2–chlorophenol adsorbed on hydrophobic clay", Environmental Science and Technology, 36(16), 3618–3624, (2002).

[396] R. Asahi, T. Morikawa, T. Ohwaki, K. Aoki, Y. Taga, "Visible–light photocatalysis in nitrogen–doped titanium oxides", Science, 293, 269–271, (2001).

[397] T. Morikawa, R. Asahi, T. Ohwaki, K. Aoki, K. Suzuki, Y. Taga, "Visible-light photocatalyst-nitrogen-doped titanium dioxide", R&D Review of Toyota CRDL, 40, 45–50, (2005).

[398] C.L. Song, T.H. Xu, Y. Liu, G. Han, "Band structures of TiO_2 doped with N, C and B", Journal of Zhejiang University Science, 7(4), 299–303, (2006).

[399] M.D. Segall, P.L.D. Lindan, M.J. Probert, C.J. Pickard, P.J. Hasnip, S.J. Clark, M.C. Payne, "First–principles simulation: ideas, illustrations and the CASTEP code", Journal of Physics: Condensed Matter, 14(11), 2717–2743, (2002).

[400] H. Irie, Y. Watanabe, K. Hashimoto, "Nitrogen–concentration dependence on photocatalytic, activity of $TiO_{2-x}N_x$ powders", Journal of Physical Chemistry B, 107(23), 5483–5486, (2003).

[401] Y. Choi, T. Umebayashi, M.Yoshikawa, "Fabrication and characterization of C–doped anatase TiO_2 photocatalysts", Journal of Materials Science, 39(5), 1837–1839, (2004).

[402] S.C. Moon, H. Mametsuka, S. Tabata, E. Suzuki, "Photocatalytic production of hydrogen from water using TiO_2 and B/TiO_2", Catalysis Today, 58(2–3),125–132, (2000).

[403] S. Sato, R. Nakamura, S. Abe, "Visible–light sensitization of TiO_2 photocatalysts by wet–method N doping", Applied Catalysis A: General, 284, 131–137, (2005).

[404] Y. Nosaka, M. Matsushita, J. Nishino, A.Y. Nosaka, "Nitrogen–doped titanium dioxide photocatalysts for visible response prepared by using organic compounds", Science and Technology of Advanced Materials, 6, 143–148, (2005).

[405] H. Qin, G. Gu, S. Liu, "Preparation of nitrogen–doped titania with visible–light activity and its application", C. R. Chimie, 11, 95–100, (2008).

[406] C. Valentin, E. Finazzi, G. Pacchioni, A. Selloni, S. Livraghi, M. C. Paganini and E. Giamello, "N–doped TiO_2: theory and experiment", Chemical Physics, 339, 44–56, (2007).

[407] J. Zhang, Y. Cong, F. Chen, M. Anpo, D. He, "Preparation, photocatalytic activity and mechanism of nano–TiO_2 co–doped with nitrogen and Iron (III)", Journal of Physical Chemistry C, 111(28), 10618–10623, (2007).

[408] C. Burda, Y. Lou, X. Chen, A.C.S. Samia, J. Stout, J.L. Gole, "Enhanced nitrogen doping in TiO_2nanoparticles", Nano Letters, 3 (8), 1049–1051, (2003).

[409] J. L. Gole, J.D. Stout, C. Burda, Y. Lou, X. Chen, "Highly efficient formation of visible light tunable $TiO_{2-x}N_x$ photocatalysts and their transformation at the nanoscale", Journal of Physical Chemistry B, 108(4), 1230–1240, (2004).

[410] S. Sakthivel, M. Janczarek, H. Kisch, "Visible light activity and photoelectrochemical properties of nitrogen–doped TiO_2", Journal of Physical Chemistry B, 108(50), 19384–19387, (2004).

[411] Y. Liu, X. Chen, J. Li, C. Burda, "Photocatalytic degradation of azo dyes by nitrogen–doped TiO_2 nanocatalysts", Chemosphere, 61, 11–18, (2005).

[412] S.M. Prokes, J.L. Gole, X. Chen, C. Burda, W.E. Carlos, "Defect–Related optical behavior in surface modified TiO_2 nanostructures", Advanced Functional Materials, 15, 161–167, (2005).

[413] C. Shifu, C. Lei, G. Shen, C. Gengyu, "The preparation of nitrogen–doped photocatalyst $TiO_{2-x}N_x$ by ball milling", Chemical Physics Letters, 413, 404–409, (2005).

[414] O. Diwald, T. L. Thompson, T. Zubkov, E. Goralski, S. Walck, J.Jr.Yates, "Photochemical activity of nitrogen–doped rutile $TiO_2(110)$ in visible light", Journal of Physical Chemistry B., 108(19), 6004–6008, (2004).

[415] R. Nakamura, T. Tanaka, Y. Nakato, "Mechanism for visible light responses in anodic photocurrents at N–Doped TiO_2 film electrodes", Journal of Physical Chemistry B, 108(30), 10617–10620, (2004).

[416] Y. Hong, C. Bang, D. Shin, H. Uhm, "Band gap narrowing of TiO_2 by nitrogen doping in atmospheric microwave plasma", Chemical Physics Letters, 413, 454–457, (2005).

[417] A. Ghicov, J.M. Macak, H. Tsuchiya, J. Kunze, V. Haeublein, L. Frey, P. Schmuki, P. "Ion implantation and annealing for an efficient N–Doping of TiO_2 nanotubes", Nano Letter, 6(5), 1080–1082, (2006).

[418] O. Diwald, T. Thompson, E. Goralski, S. Walck, T.Jr. Yates, "The effect of nitrogen ion implantation on the photoactivity of TiO_2 rutile single crystals", Journal of Physical Chemistry B, 108(1), 52–57, (2004).

[419] D.B. Hamal, K.J. Klabunde, "Synthesis, characterization and visible light activity of new nanoparticle photocatalysts based on silver, carbon and sulfur–doped TiO_2" , Journal of Colloid and Interface Science, 311, 514–522, (2007).

[420] T. Ohno, T. Tachikawa, S. Tojo, K. Kawai, M. Endo, M. Fujitsuka, K. Nishijima, Z. Miyamoto, T. Majima, "Photocatalytic oxidation reactivity of holes in the

sulfur– and carbon–doped TiO$_2$ powders studied by time–resolved diffuse reflectance spectroscopy", Journal of Physical Chemistry B,108(50), 19299–19306, (2004).

[421] T. Ohno, M. Akiyoshi, T. Umebayashi, K. Asai, T. Mitsui, M. Matsumura, "Preparation of S–doped TiO$_2$ photocatalysts and their photocatalytic activities under visible light", Applied Catalysis A: General, 265, 115–121, (2004).

[422] T. Ohno, R. Bacsa, J. Kiwi, P. Albers, V. Nadtochenko, "Preparation, testing and characterization of doped TiO$_2$ active in the peroxidation of biomolecules under visible light", Journal of Physical Chemistry B, 109(12), 5994–6003, (2005).

[423] S.Z. Chu, S. Inoue, K. Wada, D. Li, J. Suzuki, "Fabrication and photocatalytic characterizations of ordered nanoporous X–Doped (X = N, C, S, Ru, Te and Si) TiO$_2$/Al$_2$O$_3$ films on ITO/Glass", Langmuir, 21(17), 8035–8041, (2005).

[424] W. Ho, J.C. Yu, S. Lee, "Low–temperature hydrothermal synthesis of S–doped TiO$_2$ with visible light photocatalytic activity", Journal of Solid State Chemistry, 179, 1171–1176, (2006).

[425] T. Umebayashi, T. Yamaki, H. Itoh, K. Asai, "Band gap narrowing of titanium dioxide by sulfur doping", Applied Physics Letters, 81, 454, (2002).

[426] T. Umebayashi, T. Yamaki, S. Yamamoto, S. Tanaka, K. Asai, "Sulfur–doping of rutile–titanium dioxide by ion implantation: Photocurrent spectroscopy and first–principles band calculation studies", Journal of Applied Physics, 93, 5156–5160, (2003).

[427] F. Tian, C.B. Liu, "DFT description on electronic structure and optical absorption properties of anionic S–doped anatase TiO$_2$", Journal of Physical Chemistry B, 110(36), 17866–17871, (2006).

[428] W. Jin, H. Sun, Y. Bai, Y. Cheng, N. Xu, "Preparation and characterization of visible–light–driven carbon–sulfur–codoped TiO$_2$ photocatalysts", Industrial & Engineering Chemistry Research, 45(14), 4971–4976, (2006).

[429] T. Ohno, T. Mitsui, M. Matsumura, "Photocatalytic activity of S–doped TiO$_2$ photocatalyst under visible light", Chemistry Letters, 32, 364–365, (2003).

[430] V. Gombac, L. De Rogatis, A. Gasparotto, G. Vicario, T. Montini, D. Barreca, G. Balducci, P. Fornasiero, E. Tondello, M. Graziani, "TiO$_2$ nanopowders doped with boron and nitrogen for photocatalytic applications", Chemical Physics, 339, 111–123, (2007).

[431] M. Bettinelli, V. Dallacasa, D. Falcomer, P. Fornasiero, V. Gombac, T. Montini, L. Romanò, A. Speghini, "Photocatalytic activity of TiO$_2$ doped with boron and vanadium", Journal of Hazardous Materials, 146, 529–534, (2007).

[432] W. Zhao, W. Ma, C. Chen, J. Zhao, Z. Shuai, "Efficient degradation of toxic organic pollutants with Ni$_2$O$_3$/TiO$_{2-x}$B$_x$ under visible irradiation", Journal American Chemistry Society, 126(15), 4782–4783, (2004).

[433] S.U.M. Khan, M. Al–shahry, W.B. Ingler Jr, "Efficient photochemical water. splitting by a chemically modified n–TiO$_2$", Science, 297, 2243–2244, (2002).

[434] H. Irie, Y. Watanabe, K. Hashimoto, "Carbon–doped anatase TiO_2 powders as a visible–light sensitive photocatalyst", Chemistry Letters, 32, 772–773, (2003).

[435] S. Sakthivel, H. Kisch, "Daylight photocatalysis by carbon–modified titanium dioxide", Angewandte Chemie–International Edition, 42, 4908–4911, (2003).

[436] Y.Z. Li, D.S. Hwang, N.H. Lee, N.S.J. Kim, "Synthesis and characterization of carbon–doped titania as an artificial solar light sensitive photocatalyst", Chemistry and Physics Letters, 404, 25–29, (2005).

[437] X.M. Sun, Y.D. Li, "Colloidal carbon spheres and their core/shell structures with noble–metal nanoparticles", Angewandte Chemie–International Edition, 43, 597–601, (2004).

[438] L. Zhang, W. Ren, Z. Ai, F. Jia, X. Fan, Z. Zou, "Low temperature preparation and visible light photocatalytic activity of mesoporous carbon–doped crystalline TiO_2", Applied Catalysis B: Environmental, 69, 138–144, (2007).

[439] M. Shen, Z. Wu, H. Huang, Y. Du, Z. Zou, P. Yang, "Carbon–doped anatase TiO_2 obtained from TiC for photocatalysis under visible light irradiation", Materials Letters, 60, 693–697, (2006).

[440] H. Irie, S. Washizuka, K. Hashimoto, "Hydrophilicity on carbon–doped TiO_2 thin films under visible light", Thin Solid Films, 510, 21–25, (2006).

[441] J. Yang, H. Bai, X. Tan, J. Lian, "IR and XPS investigation of visible–light photocatalysis–Nitrogen–carbon–doped TiO_2 film", Applied Surface Science, 253, 1988–1994, (2006).

[442] J. Yang, H. Bai, Q. Jiang, J. Lian, "Visible–light photocatalysis in nitrogen–carbon–doped TiO_2 films obtained by heating TiO_2 gel–film in an ionized N_2 gas", Thin Solid Films, 516, 1736–1742, (2008).

[443] L. Mai, C. Huang, D. Wang, Z. Zhang, Y. Wang, "Effect of C doping on the structural and optical properties of sol–gel TiO_2 thin films", Applied Surface Science, 255, 9285–9289, (2009).

[444] D. Chen, Z. Jiang, J. Geng, Q. Wang, D. Yang, "Carbon and nitrogen co–doped TiO_2 with enhanced visible–light photocatalytic activity", Industrial & Engineering Chemistry Research, 46(9), 2741–2746, (2007).

[445] A. Hattori, M. Yamamoto, H. Tada, S. Ito, "A promoting effect of NH_4F addition on the photocatalytic activity of sol–gel TiO_2 films", Chemistry Letters, (8), 707–708, (1998).

[446] J. Yu, J.C. Yu, B. Cheng, S.K. Hark, K. Iu, "The effect of F^-–doping and temperature on the structural and textural evolution of mesoporous TiO_2 powders", Journal of Solid State Chemistry, 174, 372–380, (2003).

[447] J.C. Yu, J. Yu, W. Ho, Z. Jiang, L. Zhang, "Effects of F^- doping on the photocatalytic activity and microstructures of nanocrystalline TiO_2 powders", Chemistry of Materials, 14 (9), 3808–3816, (2002).

[448] S.N. Subbarao, Y.H. Yun, R. Kershaw, K. Dwight, A. Wold, "Comparison of the photoelectric properties of the system TiO_{2-x} with the system $TiO_{2-x}F_x$", Material Research Bulletin, 13, 1461–1467, (1978).

[449] S.N. Subbarao, Y.H. Yun, R. Kershaw, K. Dwight, A. Wold, "Electrical and optical properties of the system $TiO_{2-x}F_x$,", Inorganic Chemistry, 18(2), 488–492, (1979).

[450] D. Li, H. Haneda, S. Hishita, N. Ohashi, N.K. Labhsetwar, "Fluorine–doped TiO_2 powders prepared by spray pyrolysis and their improved photocatalytic activity for decomposition of gas–phase acetaldehyde", Journal of Fluorine Chemistry, 126, 69–77, (2005).

[451] D. Li, H. Haneda, S. Hishita, N. Ohashi, N.K. Labhsetwar, "Visible–light–driven photocatalysis on fluorine–doped TiO_2 powders by the creation of surface oxygen vacancies", Chemical Physics Letters, 401, 579–584, (2005).

[452] T. Yamaki, T. Umebayashi, T. Sumita, S. Yamamoto, M. Maekawa, A. Kawasuso, H. Itoh, "Fluorine–doping in titanium dioxide by ion implantation technique", Nuclear Instruments and Methods in Physics Research, Sect. B, 206, 254–258, (2003).

[453] Y. Su, X. Zhang, S. Han, X. Chen, L. Lei, "F–B–codoping of anodized TiO_2 nanotubes using chemical vapor deposition", Electrochemistry Communications, 9, 2291–2298, (2007).

[454] T. Giannakopoulou, N. Todorova, C. Trapalis, T. Vaimakis, "Effect of fluorine doping and SiO_2 under–layer on the optical properties of TiO_2 thin films", Materials Letters, 61, 4474–4477, (2007).

[455] D. Huang, S. Liao, J. Liu, Z. Dang, L. Petrik, "Preparation of visible–light responsive N–F–codoped TiO_2 photocatalyst by a sol–gel–solvothermal method", Journal of Photochemistry and Photobiology A: Chemistry, 184, 282–288, (2006).

[456] G. Falck, H. Lindberg, G. Falck, H. Lindberg, S. Suhonen, M. Vippola, E. Vanhala, J. Catalán, K. Savolainen, H. Norppa, "Genotoxic effects of nanosized and fine TiO_2", Human & Experimental Toxicology, 28, 339–352 (2009).

[457] S. Kim and D. Lee, "Preparation of TiO_2–coated hollow glass beads and their application to the control of algal growth in eutrophic water", Microchemical Journal, 80, 227–232, (2005).

[458] A.R. Khataee, M.N. Pons, O. Zahraa, "Photocatalytic degradation of three azo dyes using immobilized TiO_2 nanoparticles on glass plates activated by UV light irradiation: Influence of dye molecular structure", Journal of Hazardous Materials, 168, 451–457, (2009).

[459] M.J. García–Martínez, L. Canoira, G. Blázquez, I. Da Riva, R. Alcántara, J.F. Llamas, "Continuous photodegradation of naphthalene in water catalyzed by TiO_2 supported on glass Raschig rings", Chemical Engineering Journal, 110, 123–128, (2005).

[460] J. Saien, M. Asgari, A.R. Soleymani, N.Taghavinia, "Photocatalytic decomposition of direct red 16 and kinetics analysis in a conic body packed bed reactor with nanostructure titania coated Raschig rings", Chemical Engineering Journal, 151, 295–301, (2009).

[461] L. Cui, Y. Wang, M. Niu, G. Chen, Y. Cheng, "Synthesis and visible light photocatalysis of Fe–doped TiO_2 mesoporous layers deposited on hollow glass microbeads", Journal of Solid State Chemistry, 182, 2785–2790, (2009).

[462] M. Miki–Yoshida, W. Antúnez–Flores, K. Gomez–Fierro, L. Villa–Pando, R. Silveyra–Morales, P. Sánchez–Santiago, R. Martínez–Sánchez, M. José–Yacamán, "Growth and structure of TiO_2 thin films deposited inside borosilicate tubes by spray pyrolysis", Surface and Coatings Technology, 200, 4111–4116, (2006).

[463] M.C. Hidalgo, S. Sakthivel, D. Bahnemann, "Highly photoactive and stable TiO_2 coatings on sintered glass", Applied Catalysis A: General, 277, 183–189, (2004).

[464] J.M. Peralta–Hernández, J. Manríquez, Y. Meas–Vong, Francisco J. Rodríguez, T. W. Chapman, M. I. Maldonado, L. A. Godínez, "Photocatalytic properties of nano–structured TiO_2–carbon films obtained by means of electrophoretic deposition", Journal of Hazardous Materials, 147, 588–593, (2007).

[465] S. Song, T. Yang, Y. Li, Z.Y. Pang, L. Lin, M. Lv, S. Han, "Structural, electrical and optical properties of ITO films with a thin TiO_2 seed layer prepared by RF magnetron sputtering", Vacuum, 83, 1091–1094, (2009).

[466] C. He, Y. Xiong, J. Chen, C. Zha, X. Zhu, "Photoelectrochemical performance of Ag–TiO_2/ITO film and photoelectrocatalytic activity towards the oxidation of organic pollutants", Journal of Photochemistry and Photobiology A: Chemistry, 157, 71–79, (2003).

[467] H.Z. Abdullah, C.C. Sorrell, "TiO_2 thick films by electrophoretic deposition", Journal of the Australian Ceramic Society, 44, 12–16, (2008).

[468] A. Fernández, G. Lassaletta, V.M. Jiménez, A. Justo, A.R. González–Elipe, J.M. Herrmann, H. Tahiri, Y. Ait–Ichou, "Preparation and characterization of TiO_2 photocatalysts supported on various rigid supports (glass, quartz and stainless steel). Comparative studies of photocatalytic activity in water purification", Applied Catalysis B: Environmental, 7, 49–63, (1995).

[469] A. Toyoda, L. Zhang, T. Kanki, N. Sano, "Degradation of phenol in aqueous solution by TiO_2 photocatalyst coated rotating–drum reactor (SC)", Journal of Chemical Engineering of Japan, 33, 188–191, (2000).

[470] H.D. Mansilla, A. Mora, C. Pincheira, M.A. Mondaca, P.D. Marcato, N. Durán, J. Freer, "New photocatalytic reactor with TiO_2 coating on sintered glass cylinders", Applied Catalysis B: Environmental, 76, 57–63, (2007).

[471] J. Lee, M. Kim, B. Kim, "Removal of paraquat dissolved in a photoreactor with TiO_2 immobilized on the glass–tubes of UV lamps", Water Research, 36, 1776–1782, (2002).

[472] R. Scotti, M. D'Arienzo, F. Morazzoni, I. R. Bellobono, "Immobilization of hydrothermally produced TiO_2 with different phase composition for photocatalytic degradation of phenol", Applied Catalysis B: Environmental, 88, 323–330, (2009).

[473] H. Yu, S.C. Lee, J. Yu, C.H. Ao, "Photocatalytic activity of dispersed TiO$_2$ particles deposited on glass fibers", Journal of Molecular Catalysis A: Chemical, 246, 206–211, (2006).

[474] M. Khajeh Aminian, N. Taghavinia, A. Irajizad, S. Mohammad Mahdavi, "Adsorption of TiO$_2$ nanoparticles on glass fibers", Journal of Physical Chemistry C, 111, 9794–9798, (2007).

[475] S. Horikoshi, N. Watanabe, H. Onishi, H. Hidaka, N. Serpone, "Photodecomposition of a nonylphenol polyethoxylate surfactant in a cylindrical photoreactor with TiO$_2$ immobilized fiberglass cloth", Applied Catalysis B: Environmental, 37, 117–129, (2002).

[476] D. Robert, A. Piscopo, O. Heintz, J. V. Weber, "Photocatalytic detoxification with TiO$_2$ supported on glass–fibre by using artificial and natural light", Catalysis Today, 54, 291–296, (1999).

[477] V. Brezová, A. Blazková, L. Karpinský, J. Grosková, B. Havlínová, V. Jorík, M. Ceppan, "Phenol decomposition using Mn$^+$/TiO$_2$ photocatalysts supported by the sol–gel technique on glass fibres", Journal of Photochemistry and Photobiology A: Chemistry, 109, 177–183, (1997).

[478] E. Rego, J. Marto, P. Sa˜o Marcos, J.A. Labrincha, "Decolouration of Orange II solutions by TiO$_2$ and ZnO active layers screen–printed on ceramic tiles under sunlight irradiation", Applied Catalysis A: General, 355, 109–114, (2009).

[479] M. Yao , J. Chen, C. Zhao, Y. Chen, "Photocatalytic activities of Ion doped TiO$_2$ thin films when prepared on different substrates", Thin Solid Films, 517, 5994–5999, (2009).

[480] A.M. Berto, "Ceramic tiles: Above and beyond traditional applications", Journal of the European Ceramic Society, 27, 1607–1613, (2007).

[481] K. Demeestere, J. Dewulf, B.D. Witte, A. Beeldens, H. Langenhove, "Heterogeneous photocatalytic removal of toluene from air on building materials enriched with TiO$_2$", Building and Environment, 43, 406–414, (2008).

[482] S. Zhang, N. Fujii, Y. Nosaka, "The dispersion effect of TiO$_2$ loaded over ZSM–5 zeolite", Journal of Molecular Catalysis A: Chemical, 129, 219–224, (1998).

[483] J. Marto, P. S. Marcos, T. Trindade, J.A. Labrincha, "Photocatalytic decolouration of Orange II by ZnO active layers screen–printed on ceramic tiles", Journal of Hazardous Materials, 163, 36–42, (2009).

[484] P. S. Marcos, J. Marto, T. Trindade, J.A. Labrincha, "Screen–printing of TiO$_2$ photocatalytic layers on glazed ceramic tiles", Journal of Photochemistry and Photobiology A: Chemistry, 197, 125–131, (2008).

[485] G. Plesch, M. Gorbár, U. F. Vogt, K. Jesenák, M. Vargová, "Reticulated macroporous ceramic foam supported TiO$_2$ for photocatalytic applications", Materials Letters, 63, 461–463, (2009).

[486] Y. Xu, C.H. Langford, "Enhanced photoactivity of a titanium(IV) oxide supported on ZSM5 and zeolite A at low coverage", Joural of Physical Chemisty, 99, 11501–11507, (1995).

[487] Y. Xu, C.H. Langford, "Photoactivity of titanium dioxide supported on MCM41, zeolite X and zeolite Y", Joural of Physical Chemisty B, 101, 3115–3121, (1997).

[488] R.M. Mohamed, A.A. Ismail, I. Othman, I.A. Ibrahim, "Preparation of TiO_2–ZSM–5 zeolite for photodegradation of EDTA", Journal of Molecular Catalysis A: Chemical, 238, 151–157, (2005).

[489] M. Anpo, M. Takeuchi, K. Ikeue, S. Dohshi, "Design and development of titanium oxide photocatalysts operating under visible and UV light irradiation. The applications of metal ion–implantation techniques tosemiconducting TiO_2 and Ti/zeolite catalysts", Current Opinion in Solid State and Materials Science, 6, 381–388, (2002).

[490] V. Durgakumari, M. Subrahmanyam, K.V. Subba Rao, A. Ratnamala, M. Noorjahan, K. Tanaka, "An easy and efficient use of TiO_2 supported HZSM–5 and TiO_2 + HZSM–5 zeolite combinate in the photodegradation of aqueous phenol and p–chlorophenol", Applied Catalysis A: General, 234, 155–165, (2002).

[491] C.M. Zhu, L.Y. Wang, L.R. Kong, X. Yang, L.S. Wang, S.J. Zheng, F. Chen, F. M. Zhi, H. Zong, "Photocatalytic degradation of azo dyes by supported TiO_2 + UV in aqueous solution", Chemosphere, 41, 303–309, (2000).

[492] M.V. Shankar, K.K. Cheralathan, B. Arabindoo, M. Palanichamy, V. Murugesan, "Enhanced photocatalytic activity for the destruction of monocrotophos pesticide by $TiO_2/H\beta$", Journal of Molecular Catalysis A: Chemical, 223, 195–200, (2004).

[493] E.P. Reddy, L. Davydov, P. Smirniotis, "TiO_2–loaded zeolites and mesoporous materials in the sonophotocatalytic decomposition of aqueous organic pollutants: the role of the support", Applied Catalysis B: Environmental, 42, 1–11, (2003).

[494] S. Fukahori, H. Ichiura, T. Kitaoka, H. Tanaka, "Capturing of bisphenol A photodecomposition intermediates by composite TiO_2–zeolite sheets", Applied Catalysis B: Environmental, 46, 453–462, (2003).

[495] H. Chen, A. Matsumoto, N. Nisimiya, K. Tsutsumi, "Preparation and characterization of TiO_2 incorporated Y–zeolite", Colloids and Surfaces A: Physicochemical and Engineering Aspects, 157, 295–305, (1997).

[496] S. Anandan, M. Yoon, "Photocatalytic activities of the nano–sized TiO_2–supported Y–zeolites", Journal of Photochemistry and Photobiology C: Photochemistry Reviews, 4, 5–18, (2003).

[497] X.X. Wang, W.H. Lian, X.Z. Fu, J.M. Basset, F. Lefebvre, "Structure, preparation and photocatalytic activity of titanium oxides on MCM–41 surface", Journal of Catalysis, 238, 13–20, (2006).

[498] F. Li, Y. Jiang, L. Yu, Z. Yang, T. Hou, S. Sun, "Surface effect of natural zeolite (clinoptilolite) on the photocatalytic activity of TiO_2", Applied Surface Science, 252, 1410–1416, (2005).

[499] M. Huang, C. Xu, Z. Wu, Y. Huang, J. Lin, J. Wu, "Photocatalytic discolorization of methyl orange solution by Pt modified TiO_2 loaded on natural zeolite", Dyes and Pigments, 77, 327–334, (2008).

[500] H. Yamashita, Y. Ichihashi, S. Zhang, Y. Matsumura, Y. Souma, T. Tatsumi, M. Anpo, "Photocatalytic decomposition of NO at 275 K on titanium oxide catalysts anchored within zeolite cavities and framework", Applied Surface Science, 121–122, 305–309, (1997).

[501] H. Yamashita, Y. Ichihashi, M. Anpo, M. Hashimoto, C. Louis, M. Che, "Photocatalytic Decomposition of NO at 275 K on Titanium Oxides Included within Y–Zeolite Cavities: The Structure and Role of the Active Sites", Journal of Physical Chemistry, 100, 16041–16044, (1996).

[502] S. Sankararaman, K.B. Yoon, T. Yabe, J.K. Kochi, "Control of back electron transfer from charge–transfer ion pairs by zeolite supercages", Journal of the American Chemical Society, 113, 1419–1421, (1991).

[503] M. Noorjahan, V.D. Kumari, M. Subrahmanyam, P. Boule, "A novel and efficient photocatalyst: TiO_2–H–ZSM–5 combinate thin film", Applied Catalysis B: Environmental, 47, 209–213, (2004).

[504] M. Mahalakshmi, S. Vishnu Priya, Banumathi Arabindooa, M. Palanichamy, V. Murugesan, "Photocatalytic degradation of aqueous propoxur solution using TiO_2 and Hβ zeolite–supported TiO_2", Journal of Hazardous Materials, 161, 336–343, (2009).

[505] C. Wang, C. Lee, M. Lyu, L. Juang, "Photocatalytic degradation of C.I. Basic Violet 10 using TiO_2 catalysts supported by Y zeolite: An investigation of the effects of operational parameters", Dyes and Pigments, 76, 817–824, (2008).

[506] M.V. Shankar, S. Anandan, N. Venkatachalam, B. Arabindoo, V. Murugesan, "Fine route for an efficient removal of 2,4–dichlorophenoxyacetic acid (2,4–D) by zeolite–supported TiO_2", Chemosphere, 63, 1014–1021, (2006).

[507] D.I. Petkowicz, R. Brambilla, C. Radtke, C.D.S. Silva, Z.N. Rocha, S.B.C. Pergher, J.H.Z. dos Santos, "Photodegradation of methylene blue by in situ generated titania supported on a NaA zeolite", Applied Catalysis A: General, 357, 125–134, (2009).

[508] F. Li, S. Sun, Y. Jiang, M. Xia, M. Sun, B. Xue, "Photodegradation of an azo dye using immobilized nanoparticles of TiO_2 supported by natural porous mineral", Journal of Hazardous Materials, 152, 1037–1044, (2008).

[509] K. Venkata Subba Rao, A. Rachel, M. Subrahmanyam, P. Boule, "Immobilization of TiO_2 on pumice stone for the photocatalytic degradation of dyes and dye industry pollutants", Applied Catalysis B: Environmental, 46, 77–85, (2003).

[510] M. Subrahmanyam, P. Boule, V.D. Kumari, D.N. Kumar, M. Sancelme, A. Rachel, "Pumice stone supported titanium dioxide for removal of pathogen in drinking water and recalcitrant in wastewater", Solar Energy, 82, 1099–1106, (2008).

[511] A. Rachel, B. Lavedrine, M. Subrahmanyam, P. Boule, "Use of porous lavas as supports of photocatalysts", Catalysis Communications, 3, 165–171, (2002).

[512] M. Luo, D. Bowden, P. Brimblecombe, "Removal of dyes from water using a TiO_2 photocatalyst supported on black sand", Water Air Soil Pollution, 198, 233–241, (2009).

[513] R.W. Matthews, "Photooxodative degradation of coloured organics in water using supported catalysts TiO_2 and sand", Water Research, 25, 1169–1176, (1991).

[514] W. Qiu, Y. Zheng, K.A. Haralampides, "Study on a novel POM–based magnetic photocatalyst: Photocatalytic degradation and magnetic separation", Chemical Engineering Journal, 125, 165–176, (2007).

[515] S. Kurinobu, K. Tsurusaki, Y. Natui, M. Kimata, M. Hasegawa, "Decomposition of pollutants in wastewater using magnetic photocatalyst particles", Journal of Magnetism and Magnetic Materials, 310, 1025–1027, (2007).

[516] S.N. Hosseini, S.M. Borghei, M. Vossoughi, N. Taghavinia, "Immobilization of TiO_2 on perlite granules for photocatalytic degradation of phenol", Applied Catalysis B: Environmental, 74, 53–62, (2007).

[517] M. Faramarzpour, M. Vossoughi, M. Borghei, "Photocatalytic degradation of furfural by titania nanoparticles in a floating–bed photoreactor", Chemical Engineering Journal, 146, 79–85, (2009).

[518] M. Walter, M. Rüger, C. Ragor, G.C.M. Steffens, D.A. Hollander, O. Paar, H.R. Maier, W. Jahnen–Dechent, A.K. Bosserhoff, H. Erli, "In vitro behavior of a porous TiO_2/perlite composite and its surface modification with fibronectin", Biomaterials, 26, 2813–2826, (2005).

[519] M. Dogan, M. Alkan, "Adsorption kinetics of methyl violet onto perlite", Chemosphere, 50, 517–528, (2003).

[520] Y. Na, S. Song, Y. Park, "Photocatalytic decolorization of rhodamine B by immobilized TiO_2/UV in a fluidized–bed reactor", Korean Journal of Chemical Engineering, 22, 196–200, (2005).

[521] N.S. Allen, M. Edge, J. Verran, J. Stratton, J. Maltby, C. Bygott, "Photocatalytic titania based surfaces: Environmental benefits", Polymer Degradation and Stability, 93, 1632–1646, (2008).

[522] M. Lackhoff, X. Prieto, N. Nestle, F. Dehn, R. Niessner, "Photocatalytic activity of semiconductor–modified cement–influence of semiconductor type and cement ageing", Applied Catalysis B: Environmental, 43, 205–216, (2003).

[523] A. R. Khataee, A. R. Amani Ghadim, O. Valinazhad Ourang, M. Rastegar Farajzade, "Modification of white cement by titanium dioxide nanoparticles: investigation of photocatalytic activity", 2nd International Congress on Nanoscience and Nanotechnology, University of Tabriz, Tabriz, Iran, 28–30 October 2008.

[524] J. Chen, C. Poon, "Photocatalytic construction and building materials: From fundamentals to applications", Building and Environment, 44, 1899–1906, (2009).

[525] B. Neppolian, S. R. Kanel, H. C. Choi, M. V. Shankar, Banunathi Arabindoo, V. Murugesan, "Photocatalytic degradation of reactive yellow 17 dye in aqueous

solution in the presence of TiO$_2$ with cement binder", International journal of Photoenergy, 5, 45–49, (2003).

[526] L. Cassar, C. Pepe, G. Tognon, G.L. Guerrini, R. Amadelli, "White cement for architectural concrete, possessing photocatalytic properties", 11th International Congress on the Chemistry of Cement, Durban, South Africa, 11–16 May 2003, Volume 4, pp. 12.

[527] A.M. Ramirez, K. Demeestere, N. Belie, T. Mäntylä, E. Levänen, "Titanium dioxide coated cementitious materials for air purifying purposes: preparation, characterization and toluene removal potential", Building and Environment, 45, 832–838, (2010).

[528] G.L. Guerrini, F. Corazza, "White cement and photocatalysis part 1: fundamental", First Arab International Conference and Exibition on The Uses of White Cement, Cairo, Egypt, 28–30 Aprile 2008.

[529] B. Ruot, A. Plassais, F. Olive, L. Guillot, L. Bonafous, "TiO$_2$–containing cement pastes and mortars: measurements of the photocatalytic efficiency using a rhodamine B–based colourimetric test", Solar Energy, 83, 1794–1801, (2009).

[530] N. Kieda, T. Tokuhisa, "Immobilization of TiO$_2$ photocatalyst particles on stainless steel substrates by electrolytically deposited Pd and Cu", Journal of the Ceramic Society of Japan, 114, 42–45, (2006).

[531] A. Sobczyński, A. Dobosz, "Water purification by photocatalysis on semiconductors", Polish Journal of Environmental Studies, 10, 195–205, (2001).

[532] B. Kepenek, U.Ö.Ş. Seker, A.F. Cakir, M. Ürgen, C. Tamerler, "Photocatalytic bactericidal effect of TiO$_2$ thin films produced by cathodic arc deposition method", Key Engineering Materials, 254–256, 463–466, (2004).

[533] S. Yanagida, A. Nakajima, Y. Kameshima, N. Yoshida, T. Watanabe, K. Okada, "Preparation of a crack–free rough titania coating on stainless steel mesh by electrophoretic deposition", Materials Research Bulletin, 40, 1335–1344, (2005).

[534] S. Meyer, R. Gorges, G. Kreisel, "Preparation and characterisation of titanium dioxide films for catalytic applications generated by anodic spark deposition", Thin Solid Films, 450, 276–281, (2004).

[535] N. Barati, M.A. Faghihi Sani, H. Ghasemi, Z. Sadeghian, S.M.M. Mirhoseini, "Preparation of uniform TiO$_2$ nanostructure film on 316L stainless steel by sol–gel dip coating", Applied Surface Science, 255, 8328–8333, (2009).

[536] Y. Chen, D.D. Dionysiou, "TiO$_2$ photocatalytic films on stainless steel: The role of Degussa P25 in modified sol–gel methods", Applied Catalysis B: Environmental, 62, 255–264, (2006).

[537] F.D. Duminica, F. Maury, R. Hausbrand, "Growth of TiO$_2$ thin films by AP–MOCVD on stainless steel substrates for photocatalytic applications", Surface and Coatings Technology, 201, 9304–9308, (2007).

[538] P. Evans, T. English, D. Hammond, M.E. Pemble, D.W. Sheel, "The role of SiO$_2$ barrier layers in determining the structure and photocatalytic activity of TiO$_2$

films deposited on stainless steel", Applied Catalysis A: General, 321, 140–146, (2007).

[539] X.T. Zhao, K. Sakka, N. Kihara, Y. Takada, M. Arita, M. Masuda, "Structure and photo–induced features of TiO$_2$ thin films prepared by RF magnetron sputtering", Microelectronics Journal, 36, 549–551, (2005).

[540] H.Yun, J. Li, H.B. Chen, C.J. Lin, "A study on the N–, S– and Cl–modified nano–TiO$_2$ coatings for corrosion protection of stainless steel", Electrochimica Acta, 52, 6679–6685, (2007).

[541] M. Uzunova–Bujnova, R. Todorovska, M. Milanova, R. Kralchevska, D. Todorovsky, "On the spray–drying deposition of TiO$_2$ photocatalytic films", Applied Surface Science, 256, 830–837, (2009).

[542] J. Yum, S.S. Kim, D.Y. Kim, Y.E. Sung, "Electrophoretically deposited TiO$_2$ photo–electrodes for use in flexible dye–sensitized solar cells", Journal of Photochemistry and Photobiology A: Chemistry, 173, 1–6, (2005).

[543] J.I. Lim, B. Yu, K.M. Woo, Y.K. Lee, "Immobilization of TiO$_2$ nanofibers on titanium plates for implant applications", Applied Surface Science, 255, 2456–2460, (2008).

[544] H.J. Rack, J.I. Qazi, "Titanium alloys for biomedical applications", Materials Science and Engineering: C, 26, 1269–1277, (2006).

[545] B.F. Coll, P. Jacquot, "Surface modification of medical implants and surgical devices using TiN layers", Surface and Coatings Technology, 36, 867–878, (1988).

[546] M.A. Nawi, L.C. Kean, K. Tanaka, M. Sariff Jab, "Fabrication of photocatalytic TiO$_2$–epoxidized natural rubber on Al plate via electrophoretic deposition", Applied Catalysis B: Environmental, 46, 165–174, (2003).

[547] M. Lopez–Munoz, R. Grieken, J. Aguado, J. Marugan, "Role of the support on the activity of silica–supported TiO$_2$ photocatalysts: Structure of the TiO$_2$/SBA–15 photocatalysts", Catalysis Today, 101, 307–314, (2005).

[548] D. Zhao, J. Feng, Q. Huo, N. Melosh, G.H. Fredrickson, B.F. Chmelka, G.D. Stucky, "Triblock copolymer syntheses of mesoporous silica with periodic 50 to 300 angstrom pores", Science, 279, 548–552, (1998).

[549] J. Marugán, J. Aguado, W. Gernjak, S. Malato, "Solar photocatalytic degradation of dichloroacetic acid with silica–supported titania at pilot–plant scale", Catalysis Today, 129, 59–68, (2007).

[550] J. Marugán, M. López–Muñoz, J. Aguado, R. Grieken, "On the comparison of photocatalysts activity: A novel procedure for the measurement of titania surface in TiO$_2$/SiO$_2$ materials", Catalysis Today, 124, 103–109, (2007).

[551] R. Grieken, J. Marugán, C. Sordo, C. Pablos, "Comparison of the photocatalytic disinfection of *E. coli* suspensions in slurry, wall and fixed–bed reactors", Catalysis Today, 144, 48–54, (2009).

[552] J. Marugán, D. Hufschmidt, G. Sagawe, V. Selzer, D. Bahnemann, "Optical density and photonic efficiency of silica–supported TiO_2 photocatalysts", Water Research, 40, 33–839, (2006).

[553] Y.M.Wang, S.W. Liu, Z. Xiu, X.B. Jiao, X.P. Cui, J. Pan, "Preparation and photocatalytic properties of silica gel–supported TiO_2", Materials Letters, 60, 974–978, (2006).

[554] Y. Chen, K. Wang, L. Lou, "Photodegradation of dye pollutants on silica gel supported TiO_2 particles under visible light irradiation", Journal of Photochemistry and Photobiology A: Chemistry, 163, 281–287, (2004).

[555] O. Akhavan, R. Azimirad, "Photocatalytic property of Fe_2O_3 nanograin chains coated by TiO_2 nanolayer in visible light irradiation", Applied Catalysis A: General, 369, 77–82, (2009).

[556] L. Wang, H. Wang, A. Wang, M. Liu, "Surface modification of a magnetic SiO_2 support and immobilization of a Nano–TiO_2 photocatalyst on it", Chinese Journal of Catalysis, 30, 939–944, (2009).

[557] H. Li, Y. Zhang, S. Wang, Q. Wu, C. Liu, "Study on nanomagnets supported TiO_2 photocatalysts prepared by a sol–gel process in reverse microemulsion combining with solvent-thermal technique", Journal of Hazardous Materials, 169, 1045–1053, (2009).

[558] Q. He, Z. Zhang, J. Xiong, Y. Xiong, H. Xiao, "A novel biomaterial–Fe_3O_4:TiO_2 core–shell nano particle with magnetic performance and high visible light photocatalytic activity", Optical Materials, 31, 380–384, (2008).

[559] H.M. Xiao, X.M. Liu, S.Y. Fu, "Synthesis, magnetic and microwave absorbing properties of core–shell structured $MnFe_2O_4/TiO_2$ nanocomposites", Composites Science and Technology, 66, 2003–2008, (2006).

[560] H. Chen, S.W. Lee, T.H. Kim, B.Y. Hur, "Photocatalytic decomposition of benzene with plasma sprayed TiO_2–based coatings on foamed aluminum", Journal of the European Ceramic Society, 26, 2231–2239, (2006).

[561] S. Lee, K. Lee, J.H. Kim, "Titanium oxide thin film deposited on metallic Cr substrate by RF–magnetron sputtering", Materials Letters, 61, 3440–3442, (2007).

[562] L.C. Chen, F.R. Tsai, C.M. Huang, "Photocatalytic decolorization of methyl orange in aqueous medium of TiO_2 and Ag–TiO_2 immobilized on γ–Al_2O_3", Journal of Photochemistry and Photobiology A: Chemistry, 170, 7–14, (2005).

[563] C. Chis, A. Evstrator, A. Malygin, A. Malkov, P. Gaudon, J.M. Taulemeusse, "Amorphous composite photocatalysts: a new generation of active materials for environment application", Carpathian Journal of Earth And Environmental Sciences, 2, 21–28, (2007).

[564] S. Sakthivel, M. V. Shankar, M. Palanichamy, B. Arabindoo and V. Murugesan, "Photocatalytic decomposition of leather dye: Comparative study of TiO_2 supported on alumina and glass beads", Journal of Photochemistry and Photobiology A: Chemistry, 148, 153–159, (2002).

[565] B.N. Shelimov, N.N. Tolkachev, O.P. Tkachenko, G.N. Baeva, K.V. Klementiev, A.Y. Stakheev, V.B. Kazansky, "Enhancement effect of TiO_2 dispersion over alumina on the photocatalytic removal of NO_x admixtures from O_2–N_2 flow", Journal of Photochemistry and Photobiology A: Chemistry, 195, 81–88, (2008).

[566] C.J. Chung, C.C. Chiang, C.H. Chen, C.H. Hsiao, H.I. Lin, P.Y. Hsieh, J.L. He, "Photocatalytic TiO_2 on copper alloy for antimicrobial purposes", Applied Catalysis B: Environmental, 85, 103–108, (2008).

[567] J.X. Liu, D.Z. Yang, F. Shi, Y.J. Cai, "Sol–gel deposited TiO_2 film on NiTi surgical alloy for biocompatibility improvement", Thin Solid Films, 429, 225–230, (2003).

[568] T. Moskalewicz, A. Czyrska–Filemonowicz, A.R. Boccaccini, "Microstructure of nanocrystalline TiO_2 films produced by electrophoretic deposition on Ti–6Al–7Nb alloy", Surface and Coatings Technology, 201, 7467–7471, (2007).

[569] G.R. Gu, Y.A. Li, Y.C. Tao, Z. He, J.J. Li, H. Yin, W.Q. Li, Y.N. Zhao, "Investigation on the structure of TiO_2 films sputtered on alloy substrates", Vacuum, 71, 487–490, (2003).

[570] C.S. Lee, J. Kim, J.Y. Son, W. Choi, H. Kim, "Photocatalytic functional coatings of TiO_2 thin films on polymer substrate by plasma enhanced atomic layer deposition", Applied Catalysis B: Environmental, 91, 628–633, (2009).

[571] L. Zhou, S. Yan, B. Tian, J. Zhang, M. Anpo, "Preparation of TiO_2–SiO_2 film with high photocatalytic activity on PET substrate", Materials Letters, 60, 396–399, (2006).

[572] B. Sa´nchez, J.M. Coronado, R. Candal, R. Portela, I. Tejedor, M.A. Anderson, D.Tompkins, T. Lee, "Preparation of TiO_2 coatings on PET monoliths for the photocatalytic elimination of trichloroethylene in the gas phase", Applied Catalysis B: Environmental, 66, 295–301, (2006).

[573] T. Kanazawa, A. Ohmori, "Behavior of TiO_2 coating formation on PET plate by plasma spraying and evaluation of coating's photocatalytic activity", Surface & Coatings Technology, 197, 45– 50, (2005).

[574] A.H. Fostier, M.S.S. Pereira, S. Rath, J.R. Guimaraes, "Arsenic removal from water employing heterogeneous photocatalysis with TiO_2 immobilized in PET bottles", Chemosphere, 72, 319–324, (2008).

[575] Y. Zhiyong, E. Mielczarski, J. Mielczarski, D. Laub, P. Buffat, U. Klehm, P. Albers, K. Lee, A. Kulik, L. Kiwi–Minsker, A. Renken, J. Kiwi, "Preparation, stabilization and characterization of TiO_2 on thin polyethylene films (LDPE). Photocatalytic applications", Water Research, 41, 862–874, (2007).

[576] S. Naskar, S. A. Pillay, M. Chanda, "Photocatalytic degradation of organic dyes in aqueous solution with TiO_2 nanoparticles immobilized on foamed polyethylene sheet", Journal of Photochemistry and Photobiology A: Chemistry, 113, 257–264, (1998).

[577] M.P. Paschoalino, J. Kiwi, W.F. Jardim, "Gas–phase photocatalytic decontamination using polymer supported TiO$_2$", Applied Catalysis B: Environmental, 68, 68–73, (2006).

[578] J. Aarik, A. Aidla, T. Uustare, V. Sammelselg, "Morphology and structure of TiO$_2$ thin films grown by atomic layer deposition", Journal of Crystal Growth, 148, 268–275, (1995).

[579] H. Kim, H.B.R. Lee, W.J. Maeng, "Applications of atomic layer deposition to nanofabrication and emerging nanodevices", Thin Solid Films, 517, 2563–2580, (2009).

[580] M. Ritala, M. Leskela, L. Niinisto, P. Haussalo, "Titanium isopropoxide as a precursor in atomic layer epitaxy of titanium dioxide thin films", Chemistry of Materials, 5, 1174–1181, (1993).

[581] J.H. Yang, Y.S. Han, J.H. Choy, "TiO$_2$ thin–films on polymer substrates and their photocatalytic activity", Thin Solid Films, 495, 266 – 271, (2006).

[582] M.E. Fabiyi, R.L. Skelton, "Photocatalytic mineralisation of methylene blue using buoyant TiO$_2$–coated polystyrene beads", Journal of Photochemistry and Photobiology A: Chemistry, 132, 121–128, (2000).

[583] C.J. Tavares, S.M. Marques, L. Rebouta, S. L. Méndez, V. Sencadas, C.M. Costa, E. Alves, A.J. Fernandes, "PVD–Grown photocatalytic TiO$_2$ thin films on PVDF substrates for sensors and actuators applications", Thin Solid Films, 517, 1161–1166, (2008).

[584] S.X. Min, F. Wang, L. Feng, Y.C. Tong, Z.R. Yang, "Synthesis and photocatalytic activity of TiO$_2$/conjugated polymer complex nanoparticles", Chinese Chemical Letters, 19, 742–746, (2008).

[585] G. Lindbergh, E. Olsson b, B. Kasemo, M. Gustavsson, H. Ekstrom, P. Hanarp, L. Eurenius, "Thin film Pt/TiO$_2$ catalysts for the polymer electrolyte fuel cell", Journal of Power Sources, 163, 671–678, (2007).

[586] H. Yaghoubi, N. Taghavinia, E.K. Alamdari, "Self cleaning TiO$_2$ coating on polycarbonate: Surface treatment, photocatalytic and nanomechanical properties", Surface & Coatings Technology, 204, 1562–1568, (2010).

[587] J.O. Carneiro, V. Teixeira, A. Portinha, A. Magalhaes, P. Coutinho, C.J. Tavares, R. Newton, "Iron–doped photocatalytic TiO$_2$ sputtered coatings on plastics for self–cleaning applications", Materials Science and Engineering B, 138, 144–150, (2007).

[588] K. Iketani, R.D. Sun, M. Toki, K. Hirota, O. Yamaguchi, "Sol–gel–derived TiO$_2$/poly(dimethylsiloxane) hybrid films and their photocatalytic activities", Journal of Physics and Chemistry of Solids, 64, 507–513, (2003).

[589] A. Pron, P. Rannou, "Processible conjugated polymers: from organic semiconductors to organic metals and superconductors", Progress in Polymer Science, 27, 135–190, (2002).

[590] H. Liang, X. Li, "Visible–induced photocatalytic reactivity of polymer–sensitized titania nanotube films", Applied Catalysis B: Environmental, 86, 8–17, (2009).

[591] X Li, D. Wang, G. Cheng, Q. Luo, J. An, Y. Wang, "Preparation of polyaniline–modified TiO_2 nanoparticles and their photocatalytic activity under visible light illumination", Applied Catalysis B: Environmental, 81, 24, 267–273, (2008).

[592] L.X. Zhang, P. Liu, Z.X. Su, "Preparation of PANI–TiO_2 nanocomposites and their solid phase photocatalytic degradation", Polymer Degradation and Stability, 91, 2213–2219, (2006).

[593] J. Li, L. Zhu, Y. Wu, Y. Harima, A. Zhang, H. Tang, "Hybrid composites of conductive polyaniline and nanocrystalline titanium oxide prepared via self–assembling and graft polymerization", Polymer, 47, 7361–7367, (2006).

[594] D. Wang, Y. Wang, X. Li, Q. Luo, J. An, J. Yue, "Sunlight photocatalytic activity of polypyrrole–TiO_2 nanocomposites prepared by 'in situ' method", Catalysis Communications, 9, 1162–1166, (2008).

[595] D. Wang, J. Zhang, Q. Luo, X. Li, Y. Duan, J. An, "Characterization and photocatalytic activity of poly(3–hexylthiophene)–modified TiO_2 for degradation of methyl orange under visible light", Journal of Hazardous Materials, 169, 546–550, (2009).

[596] M.A. Salem, A.F. Al-Ghonemiy, A.B. Zaki, "Photocatalytic degradation of Allura red and Quinoline yellow with Polyaniline/TiO_2 nanocomposite", Applied Catalysis B: Environmental, 91, 59–66, (2009).

[597] H. Zhang, R. Zong, J. Zhao, Y. Zhu, "Dramatic visible photocatalytic degradation performances due to synergetic effect of TiO_2 with PANI", Environmental Science and Technology, 42 (10), 3803–3807, (2008).

[598] L. Song, R. Qiu, Y. Mo, D. Zhang, H. Wei, Y. Xiong, "Photodegradation of phenol in a polymer-modified TiO_2 semiconductor particulate system under the irradiation of visible light", Catalysis Communications, 8, 429–433, (2007).

[599] E. Guibal, "Heterogeneous catalysis on chitosan-based materials: a review", Progress in Polymer Science, 30, 71–109, (2005).

[600] L.F. Liu, P.H. Zhang, F.L. Yang, "Adsorptive removal of 2,4-DCP from water by fresh or regenerated chitosan/ACF/TiO_2 membrane", Separation and Purification Technology, 70, 354–361, (2010).

[601] M.S. Chiou, H.Y. Li, "Adsorption behavior of reactive dye in aqueous solution on chemical cross-linked chitosan beads", Chemosphere, 50, 1095–1105, (2003).

[602] M.S. Chiou, P.Y. Ho, H.Y. Li, "Adsorption of anionic dyes in acid solutions using chemically cross-linked chitosan beads", Dyes and Pigments, 60, 69–84, (2004).

[603] F.C. Wu, R.L. Tseng, R.S. Juang, "Enhanced abilities of highly swollen chitosan beads for color removal and tyrosinase immobilization", Journal of Hazardous Materials, 81, 167–177, (2001).

[604] C. Gerente, V. K. C. Lee, P. LE Cloirec, G. Mckay, "Application of chitosan for the removal of metals From wastewaters by adsorption–mechanisms and models review", Critical Reviews in Environmental Science and Technology, 37, 41–127, (2007).

[605] T.Y. Kim, Y.H. Lee, K.H. Park, S.J. Kim, S.Y. Cho, "A study of photocatalysis of TiO_2 coated onto chitosan beads and activated carbon", Research on Chemical Intermediates, 31, 343–358, (2005).

[606] Q. Li, H. Su, T. Tan, "Synthesis of ion-imprinted chitosan-TiO_2 adsorbent and its multi-functional performances", Biochemical Engineering Journal, 38, 212–218, (2008).

[607] C.E. Zubieta, P.V. Messina, C. Luengo, M. Dennehy, O. Pieroni, P.C. Schulz, "Reactive dyes remotion by porous TiO_2-chitosan materials", Journal of Hazardous Materials, 152, 765–777, (2008).

[608] M.M. Joshi, N.K. Labhsetwar, P.A. Mangrulkar, S.N. Tijare, S.P. Kamble, S.S. Rayalu, "Visible light induced photoreduction of methyl orange by N-doped mesoporous titania", Applied Catalysis A: General, 357, 26–33, (2009).

[609] W. Yuan, J. Ji, J. Fu, J. Shen, "A facile method to construct hybrid multilayered films as a strong and multifunctional antibacterial coating", Journal of Biomedical Materials Research Part B: Applied Biomaterials, 852, 556–563, (2007).

[610] C. Suwanchawalit, A.J. Patil, R.K. Kumar, S. Wongnawa, S. Mann, "Fabrication of ice-templated macroporous TiO_2–chitosan scaffolds for photocatalytic applications", Journal of Materials Chemistry, 19, 8478–8483, (2009).

Glossary

Activated carbon
A form of carbon that has a very high surface area (>1000 m^2/g) due to the large number of fine pores in the material. It can be regenerated (lose adsorbed gases) at room temperature.

Aerosol
A collection of solid particles or liquid droplets that are suspended in a gaseous form.

AFM: Atomic force microscope
A tip–based piso–scanning instrument able to image surfaces to molecular accuracy by mechanically probing their surface contours. It can be used for analyzing the material surface all the way down to the atoms and molecules level. A combination of mechanical and electronic probe is used in AFM to magnify surfaces up to 100,000,000 times to produce 3–D images of them.

BTXE: Benzene, Toluene, Ethylbenzene and Xylene
Toxic chemicals which are associated with petroleum products.

Carbon nanotube
A molecule first discovered in 1991 by Sumio Iijima, made from carbon atoms connected into a tube as small as 1 ϕ (nm) in diameter. It is equivalent to a flat graphene sheet rolled into a tube with high strength capacity and lightweight.

Chitosan
Chitosan is obtained by N-deacetylation of chitin, the next most abundant natural polysaccharide after cellulose. It is an invaluable renewable natural resource which technological importance is becoming increasingly evident. Chitosan is an example of basic polysaccharides. Due to this unique property many potential products using chitosan have been developed, including flocculating agents for water and wastewater treatment, chelating agents for removal of traces of heavy metals from aqueous solutions, coatings to improve dyeing characteristics of glass fibers, wet strength additives for paper, adhesives, photographic and printing applications and thickeners [599, 600].

COD: Chemical Oxygen Demand
COD is defined as the quantity of a specified oxidant that reacts with a sample under controlled conditions. The quantity of oxidant consumed is expressed in terms of its oxygen equivalence. COD is expressed in mg/L O_2. In environmental chemistry, the chemical oxygen demand (COD) test is commonly used to indirectly measure the amount of organic compounds in water. Most applications of COD determine the amount of organic pollutants found in surface water (e.g. lakes and rivers), making COD a useful measure of water quality.

Crystal structure, octahedron
An octahedron (plural: octahedra) is a polyhedron with eight faces. A regular octahedron is a Platonic solid composed of eight equilateral triangles, four of which meet at each vertex.

Crystal structure, rhombohedral
The rhombohedral (or trigonal) crystal system is one of the seven lattice point groups, named after the two–dimensional rhombus. In the rhombohedral system, the crystal is described by vectors of equal length, of which all three are not mutually orthogonal. The rhombohedral system can be thought of as the cubic system stretched diagonal along a body. a = b = c.

Crystal structure, tetragonal
A crystal structure where the axes of the unit cell are perpendicular to each other and two of the axes are of equal length but the third is not of the same length.

Crystalline (material)
A material that has a defined crystal structure where the atoms are in specific positions and are specific distances from each other.

CTAB: Cetyltrimethylammonium Bromide
Other name: Hexadecyltrimethylammonium Bromide
Chemical formula: $CH_3(CH_2)_{15}N(CH_3)_3Br$

CVD: Chemical Vapor Deposition
The deposition of atoms or molecules by the reduction or decomposition of a chemical vapor species (precursor gas) which contains the material to be deposited.

DNA: Deoxyribonucleic Acid
The carrier of genetic information, which passes from generation to generation. Every cell in the body, except red blood cells, contains a copy of the DNA.

Dopant (glass)
A chemical element that is added to give color to a glass.

Dopant (semiconductor)
A chemical element added in small amounts to a semiconductor material to establish its conductivity type and resistivity. Example: phosphorus, nitrogen, arsenic and boron.

DSSC: Dye–sensitized solar cell.

ED: Ethylenediamine $(C_2H_4(NH_2)_2)$

EDTA: Ethylene Diamine Tetraacetic Acid

Electrophoretic deposition (EPD)
Electrophoretic deposition is similar to electrochemical plating. But, instead of deposition from solution, particles are deposited from suspension. It is possible to produce thin and thick films of very consistent thickness, even on irregularly shaped substrates, with very short deposition times. Also, the equipment necessary to deposit the films has a relatively inexpensive power supply. However, the films are only physically bonded to the substrate and permanent chemical adhesion must be affected by firing, which can have deleterious results on the mechanical properties of metallic substrates.

Emulsion
An emulsion is a mixture of two immiscible (unblendable) substances. One substance (the dispersed phase) is dispersed in the other (the continuous phase). Many emulsions are oil/water emulsions (O/W), with dietary fats being one common type of oil encountered in everyday life. Examples of emulsions include butter and margarine, milk and cream. In butter and margarine, fat surrounds droplets of water (a water–in–oil emulsion). In milk and cream, water surrounds droplets of fat (an oil–in–water emulsion). See Microemulsion.

Feynman (ϕ): Nanometer (nm)
Nanoscale unit of length for the first time proposed in the present book in honor of Richard P. Feynman, the original advocate of nanoscience and nanotechnology. (One Feynman (ϕ) $\equiv$ 1 Nanometer (nm)= 10 Angstroms (Å)= 10^{-3} Micron (μ) = 10^{-9} Meter (m)).

FT–IR: Fourier Transform Infrared analysis
Infrared spectroscopy using the adsorption of infrared radiation by the molecular bonds to identify the bond types which can absorb energy by vibrating and rotating. In FT–IR the need for a mechanical slit is eliminated by frequency modulating one beam and using interferometry to choose the infrared band.

Fullerene
Fullerenes are cage–like structures of carbon atoms. There are fullerenes containing 60, 70, 80, … to 960 carbon atoms.

Hazard
A situation or condition that creates a potential exposure to something dangerous that may be harmful or injurious.

HDPM: Hexadecylpyridinium chloride–treated montmorillonite.

HeLa cell
A HeLa cell (also Hela or hela cell) is an immortal cell line used in medical research. The cell line was derived from cervical cancer cells taken from Henrietta Lacks, who died from her cancer on October 4, 1951.

HGMBs: Aluminosilicate hollow glass microbeads.

ICP: inductively coupled plasma
An inductively coupled plasma (ICP) is a type of plasma source in which the energy is supplied by electrical currents which are produced by electromagnetic induction, that is, by time–varying magnetic fields. See plasma.

In–vitro
An experimental technique performed outside a whole living organism; in a test tube.

In–vivo
An experiment performed using a living organism.

ITO: Indium–Tin Oxide.
Indium–tin oxide (ITO, or tin–doped indium oxide) is a solid miture of indium(III) oxide (In_2O_3) and tin(IV) oxide (SnO_2). It is transparent and colorless in thin layers. In bulk form, it is yellowish to grey. It is a transparent conducting material that is usually used in thin coating form. ITO is commonly used in applications such as: touch panels, electrochromic, electroluminescent and LCD displays, plasma displays, field emission displays, heat reflective coatings, energy efficient windows, gas sensors and photovoltaics.

LDPE: Low–density polyethylene film

Mesoporous and microporous material
A mesoporous material is a material containing pores with diameters between 2 and 50 nm. Porous materials are classified into several kinds by their size. According to IUPAC notation microporous materials have pore diameters of less than 2 nm and macroporous materials have pore diameters of greater than 50 nm; the mesoporous category thus lies in the middle (see J. Rouquerol, D. Avnir, C. W. Fairbridge, D. H. Everett, J. M. Haynes, N. Pernicone, J. D. F. Ramsay, K. S. W. Sing, K. K. Unger, "Recommendations for the characterization of porous solids", Pure & Applied Chemistry, 66, 1739–1758, (1994)).

Micelle and inverse micelle
A micelle is an aggregate of surfactant molecules dispersed in a liquid colloid. A typical micelle in aqueous solution forms an aggregate with the hydrophilic "head" regions in contact with surrounding solvent, sequestering the hydrophobic tail regions in the micelle centre. This type of micelle is known as a normal phase micelle (oil–in–water micelle). *Inverse micelles* have the head–groups at the centre with the tails extending out (water–in–oil micelle). Micelles are approximately spherical in shape. Other phases, including shapes such as ellipsoids, cylinders and bilayers are also possible. The shape and size of a micelle is a function of the molecular geometry of its surfactant molecules and solution conditions such as surfactant concentration, temperature, pH and ionic strength.

Microemulsions
They are clear, stable, isotropic liquid mixtures of oil, water and surfactant, frequently in combination with a cosurfactant. The aqueous phase may contain salt(s) and/or other ingredients and the "oil" may actually be a complex mixture of different hydrocarbons and olefins. In contrast to ordinary emulsions, microemulsions form upon simple mixing of the components and do not require the high shear conditions generally used in the formation of ordinary emulsions. The two basic types of microemulsions are direct (oil dispersed in water, O/W) and reversed (water dispersed in oil, W/O). See emulsion.

Nanobiotechnology
Nanotechnology applications in biological systems. Development of technology to mimic living biosystems.

Nanocatalysis
Present day production of catalysts is by tedious and expensive trial–and–error in laboratory in large–scale reactors. The catalytic action occurs on surface of highly dispersed ceramic or metallic nanostructures. Nanotechnology facilities may bring about a more scientific way of designing new catalysts named nanocatalysis with precision and predictable outcome.

Nanocomposites
Nanocomposites are materials that are created by introducing nanostructured materials (often referred to as filler) into a macroscopic sample material (often referred to as matrix). After adding nanostructured materials to the matrix material, the resulting nanocomposite may exhibit drastically enhanced properties such as electrical and thermal conductivity, optical, dielectric and mechanical properties.

Nanocrystal
Orderly crystalline aggregates of 10s–1000s of atoms or molecules with a diameter of about 10 nm.

Nanomaterial
Refers to nanoparticles, nanocrystals, nanocomposites, etc. The bottom up approach to material design.

Nanorod
Nanorods are one morphology of nanoscale objects. Each of their dimensions range from 1–100 nm.
Nanoscale
One billionth of meter scale.

Nanostructure
Geometrical structures in nanoscale.

Nanosystem
Controlled volume or controlled mass systems defined in nanoscale.

nm: nanometer = ϕ (as Feynman)

NS–TiO$_2$: nanostructured titanium dioxide.

OLEA: Oleic acid ($C_{18}H_{33}COOH$)

Photocatalysis
Photocatalysis is the acceleration of a photoreaction in the presence of a catalyst such as TiO_2.

Plasma
Plasma is an ionized gas, in which a certain proportion of electrons are free rather than being bound to an atom or molecule. The ability of the positive and negative charges to move somewhat independently makes the plasma electrically conductive so that it responds strongly to electromagnetic fields. Plasma therefore has properties quite unlike those of solids, liquids or gases. It is considered to be a distinct state of matter. Plasma typically takes the form of neutral gas–like clouds. See inductively coupled plasma (ICP).

PCD: Photocatalytic Decomposition.

PCO: Photocatalytic Oxidation.

PET: Poly Ethylene Terephthalate
PET is a thermoplastic polymer resin of the polyester family. PET is used in synthetic fibers; beverage, food and other liquid containers; thermoforming applications; and engineering resins often in combination with glass fiber.

Pumice stone
Pumice is a type of extrusive volcanic rock, produced when lava with a very high content of water and gases (together these are called *volatiles*) is extruded (or thrown out of) a volcano. As the gas bubbles escape from the lava, it becomes frothy. When this lava cools and hardens, the result is a very light rock material filled with tiny bubbles of gas. Pumice is a rock that floats on water, although it will eventually become waterlogged and sink. It is usually light–colored, indicating that it is a volcanic rock high in silica content and low in iron and magnesium, a type usually classed as *rhyolite*.

PVC: Polyvinyl chloride
It is a thermoplastic polymer which is a vinyl polymer consisting of vinyl groups (ethenyls) that are bound to chlorine.

PVD: Physical vapor deposition
The deposition of atoms or molecules that are vaporized from a solid or liquid surface. See Chemical Vapor Deposition (CVD).

PVP: Polyvinyl pyrrolidone
It is a water–soluble polymer made from the monomer N–vinylpyrrolidone.

Quantum Dots

Nanometer–sized solid state structures made of semiconductor or metal crystals capable of confining a single, or a few, electrons. The electrons possess discrete energy levels just as they would in an atom.

Salmonella choleraesuis

Salmonella choleraesuis subsp. is an important component of the porcine respiratory disease complex (PRDC). *Salmonella choleraesuis* subsp. recognized as an important and common cause of swine respiratory disease.

Self–assembly

A technique used by biological systems for assembling molecules. It is a branch of nanotechnology where objects assemble themselves with minimal external direction.

SEM: scanning electron microscope

SEM is a type of electron microscope that images the sample surface by scanning it with a high–energy beam of electrons in a raster scan pattern. The electrons interact with the atoms that make up the sample producing signals that contain information about the sample's surface topography, composition and other properties such as electrical conductivity.

Semiconductor materials

Semiconductor is a material with electrical conductivity between a good conductor and an insulator. The resistively is generally strongly temperature–dependent.

Sol–gel

The sol–gel process is a wet–chemical technique (Chemical Solution Deposition) for the fabrication of materials (typically a metal oxide) starting either from a chemical solution (sol short for solution) or colloidal particles (sol for nanoscale particle) to produce an integrated network (gel).

Sonochemistry

The study of sonochemistry is concerned with understanding the effect of sonic waves and wave properties on chemical systems.

Superhydrophilicity

Under light irradiation, water dropped onto titanium dioxide forms no contact angle (almost 0 degrees). This effect, called superhydrophilicity, was discovered in 1995 by the Research Institute of Toto Ltd. for titanium dioxide irradiated by sun light. Superhydrophilic material has various advantages. For example, it can defog glass and it can also enable oil spots to be swept away easily with water. Such materials are already commercialized as door mirrors for cars, coatings for buildings and self–cleaning surfaces.

TBAH: Tetrabutylammonium hydroxide ((C_4H_9)$_4$NOH).

TBO: Titanium(IV) butoxide (Ti(OCH$_2$CH$_2$CH$_2$CH$_3$)$_4$).

TBOT: Tetrabutyl orthotitanate (Ti(OC$_4$H$_9$)$_4$).

TEM: Transmission electron microscopy
TEM is a microscopy technique whereby a beam of electrons is transmitted through an ultra thin specimen, interacting with the specimen as it passes through it. An image is formed from the electrons transmitted through the specimen, magnified and focused by an objective lens and appears on an imaging screen. A fluorescent screen in most TEMs is detected by a sensor such as a CCD camera.

TEOA: Triethanolamine (N(CH$_2$CH$_2$OH)$_3$).

TEOS: Tetraethylorthosilicate (Si(OC$_2$H$_5$)$_4$).

TETA: Triethylenetetramine ((CH$_2$NHCH$_2$CH$_2$NH$_2$)$_2$).

TiO$_2$
Titanium dioxide, CI 77891, also known as titanium (IV) oxide, CAS No.: 13463–67–7 with molecular weight of 79.87 is the naturally occurring oxide of titanium. When used as a pigment, it is called "Titanium White" and "Pigment White 6". Titanium dioxide is extracted from a variety of naturally occurring ores that contain ilmenite, rutile, anatase and leucoxene.

TMA$^+$: Tetramethylammonium cations ((CH$_3$)$_4$N)$^+$).

TMAO: Trimethylamine–*N*–oxide dihydrate ((CH$_3$)$_3$NO·2H$_2$O).

TMD: Trimethylenediamine (H$_2$N(CH$_2$)$_3$NH$_2$).

TIPO: Titanium isopropoxide (Ti(OCH(CH$_3$)$_2$)$_4$).

TPM: *TiO$_2$* pillared montmorillonite.

TTIP: Titanium (IV) tetraisopropoxide (Ti(OCH(CH$_3$)$_2$)$_4$).
Titanium tetraisopropoxide or Titanium isopropoxide (TIPO) is a chemical compound with the formula Ti(OCH(CH$_3$)$_2$)$_4$ or (Ti(OPr)$_4$).

UV radiation: Ultraviolet radiation
Electromagnetic radiation having a wavelength in the range of 0.004 to 0.4 microns. The short wavelength UV overlaps the long wavelength Xray radiation and the long wavelengths approach the visible region.

Vapor deposition method

In general, vapor deposition methods refer to any process in which materials in a vapor state are condensed on a surface to form a solid–phase. These processes are normally used to form coatings to alter the mechanical, electrical, thermal, optical, corrosion resistance and wear resistance properties of various substrates. Recently, vapor deposition methods have been widely explored to fabricate various nanomaterials such as $NS-TiO_2$. Vapor deposition processes usually take place in a vacuum chamber. If no chemical reaction occurs, this process is called physical vapor deposition (PVD); otherwise, it is called chemical vapor deposition (CVD). In CVD processes, thermal energy heats the gases in the coating chamber and drives the deposition reaction.

XPS: X–ray photoelectron spectroscopy

XPS is a quantitative spectroscopic technique that measures the elemental composition, empirical formula, chemical state and electronic state of the elements that exist within a material. XPS spectra are obtained by irradiating a material with a beam of aluminium or magnesium X–rays while simultaneously measuring the kinetic energy (KE). XPS requires ultra–high vacuum (UHV) conditions. XPS is a surface chemical analysis technique that can be used to analyze the surface chemistry of a material in its "as received" state, or after some treatment such as: fracturing, cutting or scraping in air or UHV to expose the bulk chemistry, ion beam etching to clean off some of the surface contamination, exposure to heat to study the changes due to heating, exposure to reactive gases or solutions, exposure to ion beam implant, exposure to ultraviolet light.

XRD: X–ray diffraction

X–ray diffraction finds the geometry or shape of a molecule using X–rays. X–ray diffraction techniques are based on the elastic scattering of X–rays from structures that have long range order. The most comprehensive description of scattering from crystals is given by the dynamical theory of diffraction. XRD technique is implemented to determine crystal structure as well as crystal grain size of materials.

Zeolite

A nanoscale ceramic material that has catalytic and filtration properties.

Index

www.ingramcontent.com/pod-product-compliance
Lightning Source LLC
Chambersburg PA
CBHW050603190326
41458CB00007B/2159